AF750644

Earthworms

Earthworms: Biology and Life is a scientifically rich and comprehensive exploration of earthworm biology, diversity, and ecological significance, with a particular emphasis on the Indian subcontinent and its global context. The book begins with an overview of the global distribution and biogeography of Clitellata and Megadrili, then transitions into a detailed account of India's terrestrial and aquatic earthworm fauna. By blending regional richness with global insights, the book offers readers a nuanced understanding of species diversity, ecological roles, and habitat specialization across forest floors, agricultural lands, and freshwater ecosystems.

What sets this book apart is its integrated approach to the anatomy, physiology, and behavior of earthworms, combined with modern perspectives on their ecological importance, such as the role of gut microbiomes, vermicomposting technologies, and bioindicator functions. It synthesizes classical taxonomy with recent molecular studies, ensuring both foundational and contemporary scientific coverage. The inclusion of applied aspects like sustainable agriculture, waste management, and conservation underscores its relevance to both research and real-world applications.

Key Features

- Comprehensive coverage of earthworm taxonomy, including global distribution and Indian species diversity
- In-depth examination of life cycle, reproductive strategies, and developmental stages
- Analysis of the anatomy, behavior, and physiology related to feeding, burrowing, and reproduction
- Detailed exploration of the gut microbiome and its role in nutrient cycling and soil health
- Emphasis on ecological roles in terrestrial and aquatic environments, including invasive species impacts
- Thorough discussion on the use of earthworms as bioindicators of soil and environmental quality
- Practical insights into vermicomposting, organic waste management, and sustainable agriculture
- Evolutionary perspective linking classical taxonomy with molecular approaches
- Focus on conservation challenges such as habitat loss, pollution, and pesticide exposure
- Interdisciplinary approach combining zoology, ecology, environmental science, and applied biology

This book is an essential resource for zoologists, ecologists, soil scientists, conservationists, environmental biologists, and students in related fields. With its integration of taxonomy, physiology, ecology, and applied environmental science, *Earthworms: Biology and Life* serves both academic and practical needs, offering a holistic and authoritative perspective on one of nature's most important invertebrates. Whether for research, education, or sustainable development, this book bridges scientific insight with environmental stewardship.

Earthworms

Biology and Life

Shyamasree Ghosh, Md. Nurul Hasan, and
Chandrakanta Mandal

CRC Press
Taylor & Francis Group
Boca Raton London New York

CRC Press is an imprint of the
Taylor & Francis Group, an informa business

Designed cover image: Getty Images

First edition published 2026
by CRC Press
2385 NW Executive Center Drive, Suite 320, Boca Raton FL 33431

and by CRC Press
4 Park Square, Milton Park, Abingdon, Oxon, OX14 4RN

CRC Press is an imprint of Taylor & Francis Group, LLC

© 2026 Shyamasree Ghosh, Md. Nurul Hasan, and Chandrakanta Mandal

Reasonable efforts have been made to publish reliable data and information, but the author and publisher cannot assume responsibility for the validity of all materials or the consequences of their use. The authors and publishers have attempted to trace the copyright holders of all material reproduced in this publication and apologize to copyright holders if permission to publish in this form has not been obtained. If any copyright material has not been acknowledged please write and let us know so we may rectify in any future reprint.

Except as permitted under U.S. Copyright Law, no part of this book may be reprinted, reproduced, transmitted, or utilized in any form by any electronic, mechanical, or other means, now known or hereafter invented, including photocopying, microfilming, and recording, or in any information storage or retrieval system, without written permission from the publishers.

For permission to photocopy or use material electronically from this work, access www.copyright.com or contact the Copyright Clearance Center, Inc. (CCC), 222 Rosewood Drive, Danvers, MA 01923, 978-750-8400. For works that are not available on CCC please contact mpkbookspermissions@tandf.co.uk

Trademark notice: Product or corporate names may be trademarks or registered trademarks and are used only for identification and explanation without intent to infringe.

Library of Congress Cataloging-in-Publication Data
Names: Ghosh, Shyamasree author | Hasan, Md. Nurul 1985–author | Mandal, Chandrakanta author
Title: Earthworms : biology and life / Shyamasree Ghosh, Md. Nurul Hasan, and Chandrakanta Mandal.
Description: First edition. | Boca Raton : CRC Press, 2026. | Includes bibliographical references and index.
Identifiers: LCCN 2025035612 (print) | LCCN 2025035613 (ebook) | ISBN 9781041060772 hardback | ISBN 9781041206101 paperback | ISBN 9781003633600 ebook
Subjects: LCSH: Earthworms
Classification: LCC QL391.A6 G46 2026 (print) | LCC QL391.A6 (ebook)
LC record available at https://lccn.loc.gov/2025035612
LC ebook record available at https://lccn.loc.gov/2025035613

ISBN: 978-1-041-06077-2 (hbk)
ISBN: 978-1-041-20610-1 (pbk)
ISBN: 978-1-003-63360-0 (ebk)

DOI: 10.1201/9781003633600

Typeset in Times
by Apex CoVantage, LLC

Contents

Preface

Earthworms, often hailed as "nature's ploughmen", are fundamental to maintaining soil health, nutrient cycling, and overall ecological balance. Though they dwell quietly beneath the surface, their contributions extend far beyond the soil influencing agriculture, ecosystem dynamics, evolutionary studies, and environmental assessment.

This book, *Earthworms: Biology and Life*, offers a comprehensive exploration of the biology, diversity, ecology, and ecological significance of earthworms. It is crafted to serve as a valuable resource for students, researchers, educators, and conservation practitioners who seek to deepen their understanding of these vital invertebrates.

The volume opens with a global perspective on the geographical distribution of earthworm species and families before narrowing its focus to the diverse terrestrial and aquatic oligochaete fauna of India. It examines a range of critical topics, including habitat preferences, ecological roles, life cycles, morphology, and behavioral patterns.

Dedicated chapters explore the physiology of feeding and digestion, the complex gut microbiome, and the remarkable economic importance of earthworms in agriculture and organic waste management. The role of earthworms as bioindicators of soil health a subject of growing relevance in today's environmental context is also thoroughly addressed.

Additionally, the book considers the impact of climate change, the conservation challenges earthworms face, and the multifaceted interactions between humans and earthworm ecosystems. Insights into endemic, exotic, and semiaquatic species enrich our understanding of their natural distributions and the anthropogenic forces shaping them.

Rooted in extensive fieldwork, observation, and scholarly synthesis, this book aims not only to inform but also to inspire. It invites readers to appreciate the intricacy, resilience, and ecological importance of earthworms, humble creatures whose roles are indispensable to life beneath and above the soil.

About the Authors

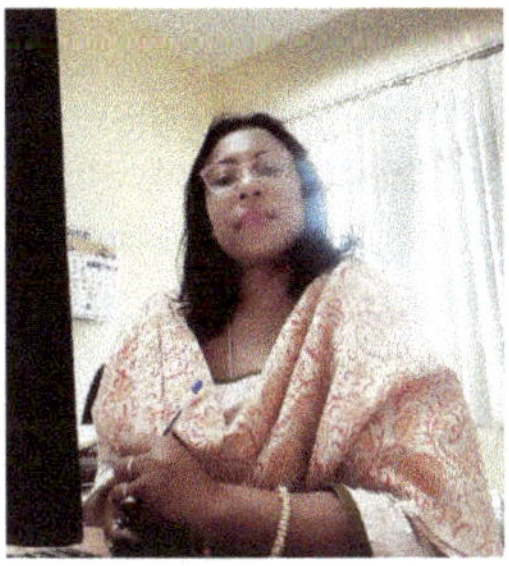

Shyamasree Ghosh MSc, PhD, PGDHE, PGDBI, has been working in the capacity of Scientific Officer in the School of Biological Sciences (SBS), National Institute of Science Education and Research (NISER), Bhubaneswar, Orissa, India, since 2009, an OCC of HBNI, Mumbai. She graduated from the prestigious Presidency College Kolkata under Calcutta University in 1998 with Honors in Zoology with Chemistry and Physiology as the other subjects. She was awarded the prestigious National Scholarship from the Government of India in the same year. She completed her post-graduation in Biotechnology from Calcutta University, ranking Second in the university in the year 2000.

She completed her PhD from the Indian Institute of Chemical Biology (IICB), CSIR, Kolkata, under Jadavpur University by qualifying for the National Eligibility Test conducted by Ministry of HRD, Govt. of India. Her PhD work in 2005 in the field of childhood acute lymphoblastic leukemia (ALL) was published in various national and international journals of repute. Later, she worked in the field of stem cells and nanotechnology in her post-doctoral research as Stem Cell Scientist in Manipal Health Systems, Bangalore, and the Indian Association for the Cultivation of Sciences (IACS), Kolkata. She served as the Head of the Biotechnology Department in a postgraduate college affiliated with Bangalore University until 2009.

She has authored and edited 16 books both as a single author and jointly with other co-authors. Her publications have been widely cited globally. She has received several awards in India for her academic endeavors.

Md. Nurul Hasan, son of Md. Mojaharul Islam and Jahanara Bibi, was born on January 3, 1985, and resides at Vill.; Hossain Pur, P.O.; Goalpara, P.S.; Chanchal, District Malda, West Bengal, India. He holds an M.Sc. in zoology and is currently pursuing a Ph.D. at the University of Calcutta. With approximately eight years of research experience, he has served in various capacities at the Zoological Survey of India, including as Subcontracted Skilled Staff (2016–2018), Junior Research Fellow (2018–2021), Senior Research Fellow (2021–2023), and currently Project Biologist (since 2024). His research interests span both invertebrate and vertebrate zoology, with specialized expertise in annelids and aquatic fauna. He has a keen focus on vermicomposting earthworm species and medicinal leeches and has conducted extensive faunal diversity surveys across India. He has been the recipient of several fellowships and honors, including the Best Research Paper Award of the Issue from the *International Journal of Advanced Research in Basic Engineering Sciences and Technology* (2016); RET qualification from the Department of Zoology, University of Calcutta (2019); and GATE

(Life Science qualification 2022). His technical skills include the collection, preservation, identification, and classification of animal specimens, along with manuscript preparation and scientific communication. He is proficient in Windows 10, MS Office, and adapts quickly to new digital tools. He is a life member of the Zoological Society, Kolkata, Indian Science Congress, and SEBA, Kolkata, and holds an ORCID ID: 0000-0002-5134-1775. He has published over 30 research articles in reputed national and international journals, contributed three abstracts, and authored ten book chapters, primarily focused on the taxonomy, biodiversity, ecology, and conservation of earthworms and leeches. He also serves as a journal referee for *Animal Biodiversity and Conservation*. He has participated in more than 20 national and international workshops and conferences, delivering oral and poster presentations as well as invited talks in webinars. Notably, he has discovered 12 new species of earthworms and leeches, significantly contributing to the advancement of zoological science in India.

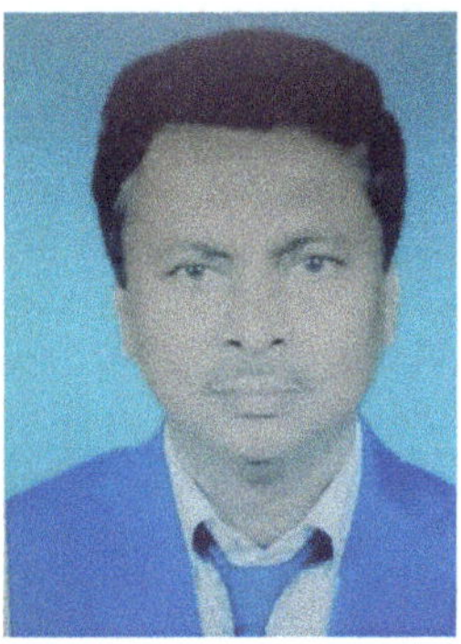

Chandrakanta Mandal, Scientist-C at the Zoological Survey of India (ZSI), Kolkata, is a distinguished zoologist with over three decades of experience in the taxonomy and ecology of annelids. He earned his Ph.D. in zoology from Kalyani University in 2004 and has served the ZSI with unwavering dedication for more than 30 years. Recognized as one of India's foremost authorities in the taxonomy, systematics, and ecological studies of earthworms and leeches, he has made profound contributions to the field of soil biodiversity. He has authored over 85 peer-reviewed research papers published in reputed national and international journals, along with numerous book chapters and scientific abstracts. A sought-after speaker, he has delivered lectures at various colleges, institutes, and universities across the country. He has a Google Scholar h-index of 10, with over 350 citations, underscoring his influential contributions to species diversity assessment, soil ecosystem studies, and the ecological role of annelids in restoration processes. He plays a critical advisory role in species identification at ZSI and provides expert support to Ph.D. scholars, academic institutions, NGOs, and research organizations across India, especially in the identification of economically and medicinally important species of earthworms and leeches. His specialized knowledge is widely applied in areas such as vermicomposting, soil health management, and the ethnomedicinal use of leeches. As a committed member of the scientific community, he serves as a reviewer for prestigious journals, including the *Records of the Zoological Survey of India, Journal of Threatened Taxa*, and the *Hippocratic Journal of Unani Medicine*. He has actively participated in over 50 seminars and 20 workshops/conferences, where he has delivered keynote presentations and training sessions. His research interests encompass taxonomy, biodiversity conservation, ecological monitoring, vermitechnology, and the therapeutic applications of annelids. His continued commitment to scientific research, environmental stewardship, and community engagement has significantly advanced zoological studies and biodiversity conservation in India.

1 Global Distribution of Earthworm Species and Families

Geographical Range of Clitellata and Megadrili

1.1 INTRODUCTION

Earthworms are essential components of terrestrial ecosystems, contributing significantly to soil health, organic matter decomposition, nutrient cycling, and overall ecosystem functioning. The global distribution of earthworm species is vast, with members of the class Clitellata being found across nearly all continents. Recent reports suggest that the worldwide distributed earthworm species 5406 species belong to 382 genera and 23 families (Mısırlıoğlu et al., 2023) and with the advances in taxonomy and molecular taxonomy, the records are being updated in a very dynamic way. These organisms are especially abundant in temperate and tropical regions, where their ecological roles are most pronounced. Earthworms belong to the suborder Megadrili, a diverse group that includes some of the largest and most ecologically significant species, particularly those found in tropical regions of South America, Southeast Asia, and Australia (Bouché, 1977).

The greatest concentration of species is in the tropical and subtropical climates, where the biodiversity of earthworms is incredibly high (Edwards & Bohlen, 1996b). Species such as *Eudrilus eugeniae* (the African nightcrawler) and *Pontoscolex corethrurus* have been widely distributed beyond their native ranges due to human activities like agriculture, with these species now found in places like Asia, Africa, South America, and the Pacific Islands (Hendrix et al., 2008). In contrast, temperate regions, such as parts of North America and Europe, are home to a wide range of earthworms, though these appear to be fewer in number compared to the tropics (Reynolds, 1994). Earthworms have become key indicators of soil health and contribute to enhancing agricultural productivity by improving soil structure and increasing the availability of nutrients to plants (Blanchart et al., 2004). Additionally, their burrowing behavior enhances soil aeration, water infiltration, and root penetration, which benefits plant growth. The movement of earthworms across regions, facilitated by trade and human activity, has led to the introduction of both beneficial and invasive species, affecting the dynamics of local ecosystems (Fragoso et al., 1999). The distribution patterns of earthworms and their ecological functions are crucial for land management, as they are integral to sustainable farming practices, soil conservation, and ecological

DOI: 10.1201/9781003633600-1

restoration efforts (Hendrix et al., 2008). These organisms not only shape the environment around them but also serve as a critical link in the food chain, providing sustenance for a variety of predators and enhancing diversity.

The available literature on earthworm species distributions at the global, regional, and national levels was reviewed and integrated with the authors' personal databases. Additionally, data from key sources were utilized, including DriloBase (2022), ITIS (2022), *Nomenclatura Oligochaetologica* (Reynolds & Wetzel, 2022), the Earthworm Species Database (Csuzdi, 2012), and Blakemore (2008) (Mısırlıoğlu et al., 2023) (Figures 1.1–1.3).

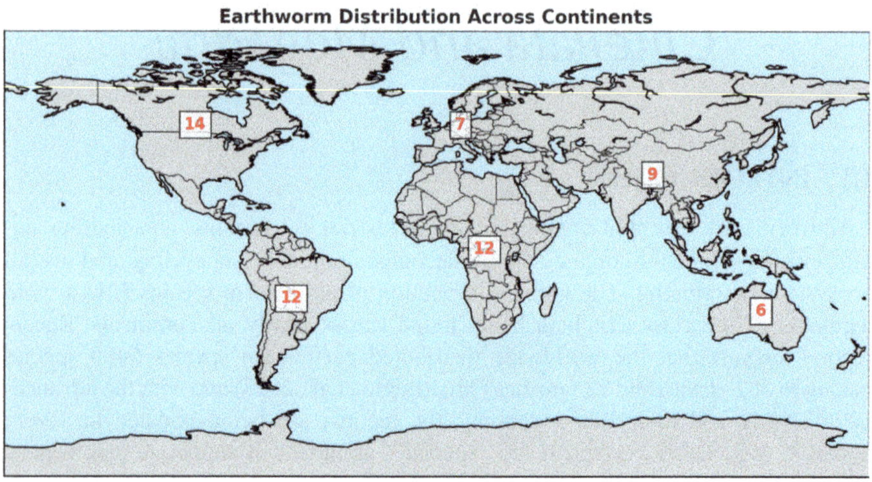

FIGURE 1.1 Map showing family distribution globally.

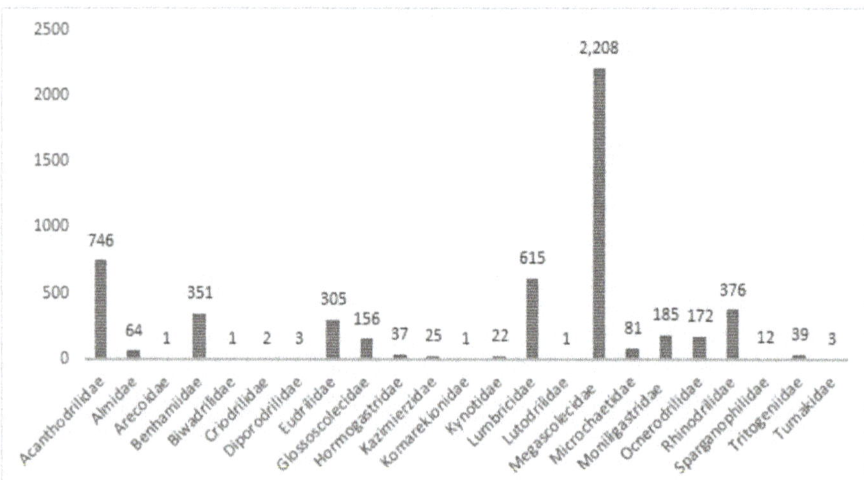

FIGURE 1.2 Worldwide family-wise earthworm diversity.

(Modified by Misirlioğlu et al., 2023).

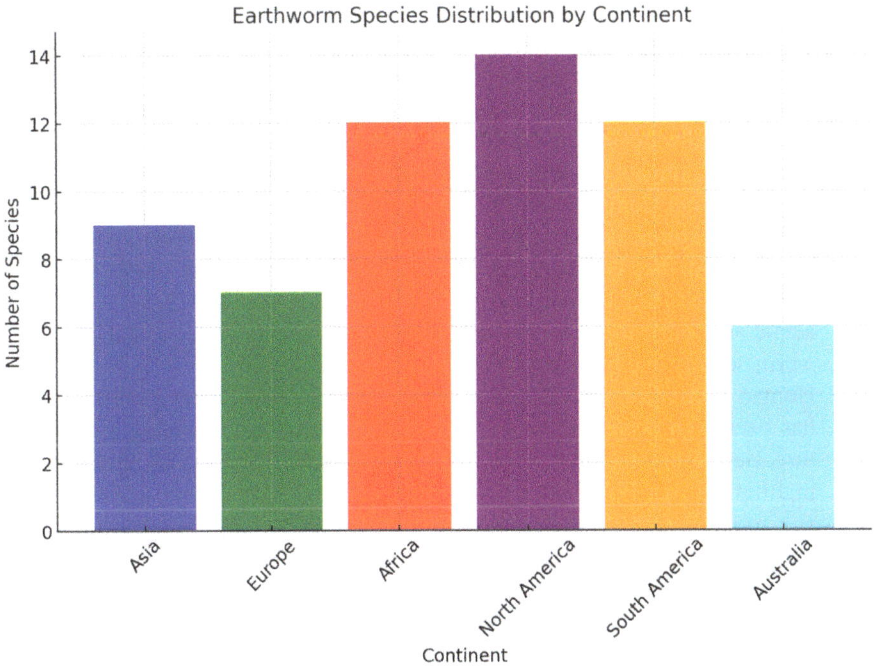

FIGURE 1.3 Earthworm family across different continents.

1.2 GLOBAL DISTRIBUTION PATTERNS AND BIOGEOGRAPHICAL REGIONS OF EARTHWORMS

Earthworms (Class: Clitellata, Order: Megadrili) exhibit diverse distribution patterns influenced by biogeographical, ecological, and anthropogenic factors. Their global distribution spans various continents, with species classified into endemic, cosmopolitan, and invasive groups. According to Fragoso & Rojas (2009a, 2009b), earthworm diversity is highest in tropical and temperate regions, with the greatest number of species found in the Neotropical, Palearctic, and Oriental regions. The primary biogeographical regions of earthworms include the Neotropical region (high diversity in the Amazon Basin), Nearctic region (introduced and native species in North America), Palearctic region (Europe, North Africa, and Asia), Afrotropical region (sub-Saharan Africa), Indomalayan region (India, Southeast Asia), Australasian region (Australia, New Zealand, and Pacific Islands), and Antarctic region, where no natural populations are reported due to extreme climatic conditions (Julka, 1993; Blakemore, 2012).

1.3 FACTORS INFLUENCING EARTHWORM DISTRIBUTION

The distribution of earthworms is governed by multiple factors, including climate, soil characteristics, and human activities.

1. Climate: Temperature and moisture are crucial in determining earthworm survival and reproduction. Lavelle et al. (2006) reported that tropical regions, with their warm and humid conditions, support high earthworm diversity, whereas extreme cold in the Arctic and Antarctic restricts their presence. Edwards & Bohlen (1996a, 1996b) emphasized that temperature fluctuations impact cocoon viability and species richness.
2. Soil Type and Composition: Different species exhibit preferences for specific soil textures, organic matter content, and pH levels. Blakemore (2018) documented that Lumbricidae species thrive in temperate, organic-rich soils, whereas tropical earthworms (e.g., *Pontoscolex corethrurus*) tolerate acidic and nutrient-poor soils. Soil porosity and aeration also affect earthworm activity and burrowing patterns (Lee, 1985).
3. Human Activities and Anthropogenic Factors: Earthworm distribution has been significantly altered by human interventions, including agriculture, deforestation, urbanization, and global trade. Hendrix et al. (2008) highlighted that European earthworm species such as *Lumbricus terrestris* and *Aporrectodea caliginosa* have been introduced to North America and other regions due to soil transport, leading to ecological imbalances. Deforestation in tropical regions alters soil moisture and organic matter availability, affecting endemic species (Brown et al., 2006b). Additionally, climate change may shift earthworm distributions, as rising temperatures enable species to expand their ranges into previously unsuitable habitats (Eisenhauer et al., 2014).

1.4 CLIMATE CHANGE AND DISTRIBUTION SHIFTS IN EARTHWORMS

Climate change is a significant driver of ecological shifts, and earthworm populations are no exception. Rising global temperatures, altered precipitation patterns, and extreme weather events directly influence earthworm survival, reproduction, and distribution. Earthworms are poikilothermic organisms, meaning their physiological activities are highly dependent on external temperature and moisture conditions (Edwards & Bohlen, 1996a, 1996b). A warming climate may allow some species to expand their ranges, while others face population declines due to habitat unsuitability.

1.4.1 Temperature and Moisture Effects

Temperature changes alter earthworm metabolic rates, growth, and reproductive success. Eisenhauer et al. (2014) noted that increasing temperatures could benefit temperate earthworm species like *Lumbricus terrestris* by extending their active season, leading to greater reproduction and dispersal. However, tropical species that are adapted to stable temperature ranges may face thermal stress and decline in population numbers (Lavelle et al., 2006). Additionally, increased evaporation due to warming can lead to soil desiccation, reducing earthworm activity and survival in arid and semi-arid regions (Blouin et al., 2013).

1.4.2 Shifts in Precipitation and Extreme Weather

Changes in rainfall patterns are another crucial factor. Marhan et al. (2015) found that prolonged droughts negatively impact anecic earthworms (deep burrowers), forcing them into aestivation, which limits their feeding and reproduction. Conversely, excessive rainfall may lead to waterlogged soils, reducing oxygen availability and causing earthworm mortality due to anoxia (Bartz et al., 2013). Flood-prone regions may see reduced earthworm populations, while areas with moderate increases in moisture may experience higher earthworm densities.

1.4.3 Predictive Models for Future Distribution

Several species distribution models (SDMs) and ecological niche models (ENMs) have been used to predict the future distribution of earthworms under different climate change scenarios. Phillips et al. (2019) utilized MaxEnt modeling to predict the potential shifts of *Lumbricus rubellus* under various climate conditions, showing northward expansion in response to warming in temperate regions. Similarly, Eggleton et al. (2009) applied bioclimatic envelope models to forecast how earthworm biodiversity hotspots might shift due to increasing temperatures. The results indicated that by 2100, several European and North American species might migrate poleward as they track suitable climate conditions, while equatorial regions could suffer a loss of endemic species.

1.4.4 Long-Term Ecological Impacts and Conservation Strategies

The redistribution of earthworm populations duc to climate change has cascading effects on soil health, carbon sequestration, and nutrient cycling (Ferlian et al., 2020). The expansion of certain invasive earthworm species into previously uncolonized regions, such as the northern boreal forests, can disrupt soil microbial communities and accelerate organic matter decomposition (Gundale et al., 2005). Conservation efforts must focus on monitoring climate-vulnerable species, maintaining habitat connectivity, and integrating earthworm biodiversity into climate action frameworks (Eisenhauer & Straube, 2021).

Climate change is reshaping global earthworm distributions, with temperature increases and altered precipitation patterns playing a crucial role. Predictive models indicate that while some species will expand their ranges, others will experience range contractions or localized extinctions. Future research should refine distribution models by incorporating microhabitat preferences, soil properties, and genetic adaptability to climate stress. These insights will be vital for conservation planning and ecosystem management in a rapidly changing world. We discuss the impact of climate change on earthworms in subsequent chapters.

1.5 GLOBAL DISTRIBUTION OF EARTHWORMS

Earthworms exhibit a diverse range of species compositions and ecological adaptations across different continents. With the increasing application of modern molecular biology tools in taxonomy, and advances of our knowledge, the list of biodiversity of earthworm is growing. Their distribution is influenced by climatic conditions, soil types, and historical biogeographic factors. Following is a detailed account of their presence in various regions, including major species, families, biodiversity hotspots, and their ecological roles.

1.5.1 Asia: Major Species and Families, Biodiversity Hotspots

Asia harbors a high diversity of earthworms due to its varied climatic zones, from temperate forests to tropical rainforests and arid regions. The dominant families include Megascolecidae, Moniligastridae, and Lumbricidae (Blakemore, 2006). The Western Ghats, Himalayas, and Indo-Burma regions are recognized as major biodiversity hotspots for earthworms (Julka & Paliwal, 1993). Species such as *Drawida japonica* (Moniligastridae) and *Amynthas cortices* (Megascolecidae) are common in these regions (Shen et al., 2011). Agricultural intensification and deforestation have significantly impacted native earthworm populations, leading to habitat fragmentation and biodiversity loss (Tripathi & Bhardwaj, 2004).

1.5.2 Europe: Native and Invasive Species, Soil Impact

Europe is home to both native and introduced earthworm species, with the Lumbricidae family dominating the continent. Native species like *Lumbricus terrestris* and *Aporrectodea caliginosa* play crucial roles in soil aeration and organic matter decomposition (Edwards & Bohlen, 1996a, 1996b). However, invasive species such as *Eisenia fetida* and *Dendrobaena veneta* introduced through agriculture and composting have altered soil microbial communities and nutrient cycles (Curry & Schmidt, 2007). The invasion of European species into North America, particularly in previously glaciated regions, has led to significant changes in forest soil structure and carbon storage (Hendrix et al., 2008).

1.5.3 Africa: Tropical and Arid-Region Adaptations

Africa exhibits distinct earthworm fauna adapted to both humid tropical forests and arid savannas. The dominant families include Eudrilidae and Acanthodrilidae (Decaëns et al., 2006). Tropical species such as *Eudrilus eugeniae* thrive in warm, moist conditions and are widely used in vermiculture and organic waste processing (Lavelle et al., 1998). In contrast, arid-adapted species like *Microchaetus rappi* have evolved burrowing behaviors to survive prolonged droughts (Plisko, 2010). Soil degradation, desertification, and land-use changes have threatened native earthworm populations in several African regions (Ayuke et al., 2011).

1.5.4 North America: Introduced vs. Native Species

North America features a mix of native and introduced earthworm species, with the latter significantly impacting native ecosystems. The Lumbricidae family dominates, with species such as *Lumbricus rubellus* and *Aporrectodea trapezoides* being widespread (Reynolds, 2008a, 2008b). The northern forests, which lacked earthworms since the last Ice Age, have been dramatically altered by the invasion of European species (Bohlen et al., 2004a, 2004b). These introduced earthworms accelerate leaf litter decomposition, disrupting native plant communities and altering nutrient cycling (Frelich et al., 2006). In contrast, native species like *Diplocardia ornata* in southern states contribute to natural soil fertility and ecosystem stability (Callaham et al., 2002).

1.5.5 South America: Amazon Basin and Highland Diversity

South America hosts a diverse array of earthworms, particularly in the Amazon rainforest and Andean highlands. The dominant families include Glossoscolecidae and Rhinodrilidae (Fragoso & Brown, 2007a, 2007b). Species such as *Pontoscolex corethrurus* are highly adaptable and thrive in both disturbed and undisturbed ecosystems (Decaëns et al., 1999). In the Andean regions, endemic species have developed cold tolerance and unique burrowing behaviors to survive in high-altitude environments (Lavelle & Lapied, 2003). However, deforestation and agricultural expansion are threatening several native species, particularly those with restricted ranges (Fragoso et al., 1997a, 1999b).

1.5.6 Australia and Oceania: Unique Species and Adaptations

Australia and Oceania are home to unique earthworm species, many of which are megadriles (large-bodied earthworms). The Megascolecidae family dominates, with species such as *Megascolides australis*, a giant earthworm that can reach over 3 meters in length (Dyne, 1979, 1997). New Zealand's native earthworm fauna includes species adapted to acidic and volcanic soils, such as *Octochaetus multiporus* (Lee, 1959). However, invasive European species have outcompeted many native earthworms, leading to shifts in soil ecology and organic matter decomposition (Baker et al., 2006).

1.6 BIOGEOGRAPHICAL PATTERNS OF EARTHWORMS

The distribution of earthworms across the globe has been shaped by historical geological events, particularly the breakup of supercontinents such as Pangaea, Gondwana, and Laurasia. These events influenced the dispersal, speciation, and adaptation of earthworm lineages, leading to the distinct biogeographical patterns observed today.

1.7 ROLE OF HISTORICAL GEOLOGY IN SPECIES DISPERSION

Earthworm distribution is closely linked to plate tectonics, continental drift, and climatic shifts over geological time. Fossil records and molecular phylogenetics suggest that early earthworms evolved before the breakup of Pangaea, with subsequent continental separations leading to isolated evolutionary lineages (James & Davidson, 2012). As continents drifted apart, earthworm populations were subjected to new environmental pressures, resulting in the diversification of families and genera across different landmasses (Reynolds, 1994, Table 1.1).

TABLE 1.1
Major Historical Geological Events That Influenced Earthworm Distribution

Breakup of Pangaea (~200 Mya)	Initial divergence of earthworm ancestors into different supercontinents
Separation of Laurasia and Gondwana (~180 Mya)	Formation of distinct earthworm lineages in the Northern and Southern Hemispheres
Further fragmentation of Gondwana (~100–50 Mya)	Isolation of species in South America, Africa, Australia, and India, leading to regional endemism
Quaternary Glaciations (~2.5 Mya–12,000 years ago)	Extirpation of earthworms in glaciated regions and post-glacial recolonization by invasive species

TABLE 1.2
Origins and Distributions of Earthworms

South America	High diversity of Glossoscolecidae in the Amazon Basin, adapted to humid tropical conditions (Fragoso & Brown, 2007a, 2007b)
Africa	Presence of Eudrilidae, specialized for arid and semi-arid environments (Plisko, 2010)
Australia	Evolution of giant Megascolecidae, such as *Megascolides australis*, due to stable and nutrient-poor soils (Dyne, 1997 Evolution of giant Megascolecidae, such as *Megascolides australis*, due to stable and nutrient-poor soils (Dyne, 1997)
India	High endemism of Octochaetidae, reflecting its prolonged geological isolation before colliding with Asia (Julka, 1993). High endemism of Octochaetidae, reflecting its prolonged geological isolation before colliding with Asia (Julka, 1993)

1.7.1 Pangean Origins of Earthworms

During the existence of Pangaea (335–200 Mya), earthworms were likely present in the moist, warm climates of the supercontinent (Blakemore, 2012). Fossil and molecular data suggest that ancestral Clitellata diversified early, with primitive families such as Moniligastridae and Lutodrilidae appearing before continental drift (Chang & Chen, 2005). When Pangaea split into Laurasia and Gondwana, these ancestral groups were divided, forming the foundation for modern biogeographical patterns.

1.8 GONDWANAN ORIGINS AND DISTRIBUTION

The Gondwanan landmasses (South America, Africa, Australia, Antarctica, and India) share many endemic earthworm families, such as Acanthodrilidae, Glossoscolecidae, and Megascolecidae (Jamieson et al., 2002). These families diversified in response to regional climatic and geological changes as in Table 1.2:

1.9 LAURASIAN ORIGINS AND DISTRIBUTION

Laurasia, which included North America, Europe, and parts of Asia, was dominated by earthworms from the Lumbricidae family. These species adapted to temperate climates and played a crucial role in soil formation during the early Cenozoic. The ice ages led to the loss of native earthworms in many glaciated regions, particularly in North America, where post-glacial recolonization was largely driven by European Lumbricidae invasions (Hendrix et al., 2008).

1.10 EARTHWORMS IN SOIL ECOSYSTEMS

Earthworms are also called "ecosystem engineers" due to their influence on soil structure, fertility, and biological processes, and they are known to play a critical role in nutrient cycling, organic matter decomposition, and soil aeration, making them indispensable in both natural and agricultural ecosystems.

Their burrowing activities enhance water infiltration, root penetration, and microbial interactions, improving overall soil health. The impact of earthworms on agriculture and forestry varies globally, depending on the climate, soil type, and species composition.

1.10.1 Role in Nutrient Cycling and Soil Fertility

Earthworms contribute to nutrient cycling by breaking down organic matter, mineralizing nutrients, and redistributing soil components. Their activities influence key soil nutrients such as nitrogen (N), phosphorus (P), and potassium (K), which are essential for plant growth (Edwards & Bohlen, 1996a, 1996b).

The process of vermicomposting, where earthworms digest organic residues and excrete nutrient-rich casts, is an important mechanism for enhancing soil fertility.

1.10.2 Decomposition of Organic Matter

Earthworms consume dead plant material and soil microorganisms, accelerating the breakdown of organic matter. This releases essential nutrients in forms readily available for plant uptake (Lavelle et al., 1998). Their gut microbiota plays a crucial role in decomposing complex organic compounds, aiding in carbon and nitrogen cycling (Blouin et al., 2013).

1.10.3 Formation of Soil Aggregates

Earthworm casts are rich in minerals and organic matter, which improve aggregation and stability of soil (Six et al., 2002). Aggregation prevents soil erosion and enhances water retention, making soils more resilient to drought and heavy rainfall events.

1.10.4 Enhancement of Microbial Activity

The mucus secreted by earthworms during burrowing promotes microbial growth, increasing enzymatic activity and nutrient mineralization rates (Aira et al., 2008). Their interactions with bacteria and fungi enhance nitrogen fixation, benefiting crops and natural vegetation.

1.10.5 Carbon Sequestration

Earthworm activity influences soil carbon storage and greenhouse gas emissions. By increasing decomposition rates, they accelerate carbon cycling, but they can also enhance long-term carbon sequestration by stabilizing organic matter in deep soil layers (Fierer et al., 2009).

1.10.6 Impact on Agriculture and Forestry Globally

1.10.6.1 Agriculture

Earthworms play an important role in agriculture by improving soil fertility, reducing the use of chemical fertilizers, and enhancing crop productivity. However, their impact varies depending on soil type, climate, and farming practices.

- Positive Effects on Crop Yields: Studies show that earthworm activity can increase crop productivity by 25–30% due to enhanced soil aeration, root penetration, and nutrient cycling (van Groenigen et al., 2014). Their presence is particularly beneficial in organic farming and no-till agriculture, where they help incorporate organic amendments into the soil (Baker et al., 1999).
- Improved Soil Porosity and Water Retention: Earthworm burrowing creates macropores, improving soil aeration and water infiltration. This reduces waterlogging in clay-rich soils and improves moisture retention in sandy soils (Edwards & Arancon, 2004).
- Biological Pest and Disease Control: Earthworms contribute to soil disease suppression by enhancing microbial diversity, which can inhibit plant pathogens such as Fusarium and Pythium (Blouin et al., 2013). Their casts also contain antimicrobial compounds that reduce soil-borne diseases.
- Potential Negative Effects in Certain Conditions: While earthworms are generally beneficial, their activity can be problematic in intensive agriculture. In some cases, their burrowing and mixing of soil layers may increase nitrate leaching, leading to groundwater contamination (Lubbers et al., 2017). Additionally, some invasive species, such as *Amynthas agrestis* (Asian jumping worms), disrupt soil structure and reduce organic matter stability (Görres et al., 2019).

1.10.6.2 Forestry

The role of earthworms in forestry depends on the ecosystem type and dominant tree species. In temperate and tropical forests, earthworms enhance decomposition, soil aeration, and nutrient cycling, promoting tree growth. However, in glaciated regions such as North America, invasive earthworms have disrupted forest floor stability, leading to ecological imbalances (Bohlen et al., 2004a, 2004b).

- Tropical Forests: In tropical regions, native earthworms contribute to soil fertility and reforestation efforts by breaking down leaf litter and improving nutrient cycling (Fragoso et al., 1997b). In agroforestry systems, their activity enhances root development and stabilizes tree seedlings.

- Boreal and Temperate Forests: In North America and parts of Europe, invasive European earthworms have significantly altered forest soil composition, leading to declines in native plant species and increased soil erosion (Frelich et al., 2006). Some tree species, such as oak and maple, struggle to regenerate in heavily invaded soils.
- Afforestation and Soil Restoration: Earthworm-mediated soil improvement is beneficial in afforestation and land reclamation projects, especially in degraded lands. Vermicomposting and biochar applications combined with earthworm activity have shown promise in restoring soil fertility in deforested areas (Hendrix et al., 2008).

1.11 CONSERVATION AND THREATS TO EARTHWORM POPULATIONS

Although earthworms play a vital role in ecosystem functioning, particularly in soil health, organic matter decomposition, and nutrient cycling, their populations are increasingly threatened by impacts of destruction of habitat, climate change, pollution, and invasive species. Conservation strategies, including habitat protection, sustainable agriculture, and species monitoring, are crucial for maintaining earthworm biodiversity worldwide.

1.12 HABITAT DESTRUCTION AND POLLUTION IMPACT ON EARTHWORMS

1.12.1 Deforestation and Land-Use Changes

Large-scale deforestation and agricultural expansion significantly impact earthworm habitats. The removal of forest cover and topsoil leads to loss of organic matter, reducing food availability for earthworms (Fragoso et al., 1997b). Urbanization and infrastructure development further contribute to habitat fragmentation, restricting earthworm dispersal and genetic diversity (Bohlen et al., 2004a, 2004b).

1.12.2 Intensive Agriculture and Pesticides

The widespread use of chemical fertilizers, pesticides, and herbicides negatively affects earthworm populations. Studies have shown that organophosphate pesticides and neonicotinoids reduce earthworm reproduction and survival rates (Pelosi et al., 2013). Tillage and soil compaction from mechanized farming disrupt earthworm burrows, leading to population declines in intensively cultivated lands (Edwards & Bohlen, 1996a, 1996b).

1.12.3 Soil Pollution and Heavy Metals

Industrial pollution and mining activities have led to soil contamination with heavy metals (e.g., lead, cadmium, mercury), which are toxic to earthworms. These pollutants accumulate in their tissues, affecting their metabolism and reproduction

(Spurgeon & Hopkin, 1996). Earthworms play a role in bioremediation, but excessive pollution reduces their effectiveness and leads to population collapse in contaminated areas (Nahmani et al., 2007).

1.12.4 Climate Change and Extreme Weather Events

Changes in temperature and precipitation patterns are altering earthworm distribution and behavior. Prolonged droughts reduce soil moisture, making it difficult for earthworms to survive, while heavy rainfall can lead to waterlogged soils, suffocating them (Blouin et al., 2013). In colder regions, warming temperatures may allow invasive species to expand, outcompeting native earthworms (Lubbers et al., 2017).

1.12.5 Invasive Earthworm Species

The introduction of non-native earthworm species can disrupt native ecosystems. In North America, European species (*Lumbricus terrestris*, *Aporrectodea* spp.) have altered soil structure and nutrient cycles that are known to adversely impact the native plant communities (Hendrix et al., 2008). Similarly, invasive Amynthas (jumping worms) degrade organic matter too quickly, leading to nutrient depletion and soil erosion (Görres et al., 2019).

1.13 STRATEGIES FOR ENDEMIC EARTHWORM SPECIES GLOBALLY

1. Habitat Protection and Restoration
   - Conservation of forests, grasslands, and wetlands is crucial for maintaining earthworm biodiversity. Protected areas including wildlife reserves and national parks serve as safe zones for endemic species (Fragoso et al., 1997a, 1999).
   - Reforestation and agroforestry programs incorporating organic farming principles can restore earthworm populations in degraded landscapes (Edwards & Arancon, 2004).
2. Sustainable Agricultural Practices
   - This is being implemented by reducing use of pesticide and fertilizer through integrated pest management (IPM), and organic farming helps confer protection to earthworms.
   - Conservation tillage, cover cropping, and crop rotation improve soil structure and provide suitable habitats for earthworms (van Groenigen et al., 2014).
3. Pollution Control and Soil Remediation
   - Bioremediation strategies, including vermicomposting and biochar applications, help mitigate soil contamination while supporting earthworm populations (Sinha et al., 2010).
   - Regulations on industrial waste disposal and mining activities are needed to prevent toxic heavy metal accumulation in soils (Nahmani et al., 2007).

4. Invasive Species Management
    - Efforts to monitor and control invasive earthworm species should be prioritized in sensitive ecosystems, particularly in boreal forests where native species are at risk (Frelich et al., 2006).
    - Research on biological control methods and habitat restoration could help mitigate the impact of invasive species without harming native earthworms.
5. Climate Adaptation and Monitoring Programs
    - Long-term monitoring programs are essential to track earthworm population trends in response to climate change (Lubbers et al., 2017).
    - Conservation efforts should incorporate climate adaptation strategies, such as creating microhabitats with stable moisture levels to support earthworm survival during extreme weather conditions.
6. Public Awareness and Policy Implementation
    - Raising awareness about the importance of earthworms in soil health and food security is crucial for conservation.
    - Governments should incorporate soil biodiversity protection policies into land-use planning and sustainable development goals (Hendrix & Bohlen, 2002).

1.14 CONSERVATION AND THREATS TO EARTHWORMS: A GLOBAL PERSPECTIVE

Earthworms, as key ecosystem engineers, play an essential role in soil health, organic matter decomposition, and nutrient cycling. However, their populations are increasingly under threat due to destruction of habitat, pollution, invasive species, and impact of climate change. The loss of native earthworm species has cascading effects on soil biodiversity, microbial communities, and plant growth, which in turn affect agriculture, forestry, and natural ecosystems (Edwards & Bohlen, 1996a, 1996b). To mitigate these threats, global conservation strategies must focus on habitat restoration, pollution control, sustainable land management, and monitoring programs.

1.14.1 Habitat Destruction and Its Impact on Earthworms

One of the significant threats to earthworm populations globally is habitat destruction, which occurs due to deforestation, intensive agriculture, urbanization, and mining activities. Soil compaction, loss of organic matter, and reduced moisture levels directly affect earthworm survival and reproduction (Fragoso et al., 1997b).

- Deforestation and Land Conversion: Large-scale deforestation in tropical regions like the Amazon Basin, Southeast Asia, and Central Africa has led to a decline in native earthworm populations by removing the organic leaf litter and soil moisture essential for their survival (Decaëns et al., 2006).
- Intensive Agriculture: Conventional farming practices, including deep plowing, monoculture plantations, and excessive chemical inputs, reduce

earthworm abundance by destroying soil structure and food sources (van Groenigen et al., 2014).
- Urbanization and Infrastructure Development: Expanding urban centers, roads, and industrial areas lead to soil sealing and habitat fragmentation, limiting earthworm movement and genetic diversity (Bohlen et al., 2004a, 2004b).
- Mining and Land Degradation: Mining activities expose subsoil layers, making them inhospitable for earthworms. Heavy machinery compacts the soil, further reducing earthworm populations in affected areas (Nahmani et al., 2007).

1.14.2 Pollution and Its Effects on Earthworms

Pollution is another major factor affecting earthworm populations. Soil contamination with heavy metals, pesticides, and industrial waste can lead to toxicity, reproductive failure, and population decline (Spurgeon & Hopkin, 1996).

- Pesticide and Fertilizer Use: Many chemical pesticides, such as neonicotinoids and organophosphates, have been shown to reduce earthworm reproduction and disrupt their burrowing activities (Pelosi et al., 2013). High levels of nitrogen-based fertilizers can alter soil pH, affecting earthworm diversity (Edwards & Arancon, 2004).
- Heavy Metal Contamination: Mining, industrial waste disposal, and vehicular emissions contribute to lead (Pb), cadmium (Cd), and mercury (Hg) contamination, which bioaccumulates in earthworms, reducing their metabolism and survival rates (Nahmani et al., 2007).
- Plastic and Microplastic Pollution: Recent studies have shown that microplastics alter soil structure and negatively impact earthworm feeding behavior and reproduction (Rodriguez-Seijo et al., 2017).

1.14.3 Conservation Strategies for Endemic Earthworm Species

Earthworm conservation requires multi-faceted global approaches, focusing on habitat protection, sustainable agriculture, pollution reduction, and climate adaptation.

1. Habitat Protection and Restoration
   - Establishing protected areas for native earthworms in tropical rainforests, grasslands, and wetlands can help preserve endemic species. For example, conservation efforts in Amazonian forests have successfully maintained diverse earthworm communities (Fragoso et al., 1997a, 1999).
   - Agroforestry systems and organic farming provide suitable habitats for earthworms by reducing tillage, increasing organic matter inputs, and minimizing chemical use (Lubbers et al., 2017).
   - Soil conservation programs that promote reforestation and sustainable land-use practices can help restore degraded habitats.

2. Sustainable Agriculture and Land Management
   - Reducing tillage intensity and monoculture plantations improves soil aeration and enhances earthworm populations (Blouin et al., 2013).
   - Incorporating vermicomposting and organic fertilizers instead of chemical inputs supports healthy earthworm communities (Sinha et al., 2010).
   - Adopting no-till farming and cover cropping improves soil organic matter retention and moisture, benefiting earthworm biodiversity (van Groenigen et al., 2014).
3. Pollution Control and Soil Remediation
   - Bioremediation strategies, such as using earthworms in vermiculture and soil detoxification, can mitigate heavy metal contamination (Nahmani et al., 2007).
   - Implementing strict regulations on industrial waste disposal and reducing pesticide use can prevent soil degradation.
4. Climate Adaptation and Monitoring
   - Long-term monitoring programs tracking earthworm distribution and responses to climate change are crucial for conservation planning (Lubbers et al., 2017).
   - Creating microhabitats and moisture-retaining soil structures in arid and semi-arid regions can help earthworms adapt to changing climates.
5. Addressing Invasive Species
   - Controlling invasive earthworms, such as *Amynthas* spp. (Asian jumping worms) in North America, can prevent soil degradation and protect native earthworm populations (Görres et al., 2019).
   - Encouraging research on biological control and native species restoration can help mitigate the effects of non-native earthworms.

1.14.4 Global Policy Recommendations and Public Awareness

- International Collaboration: Governments, conservation organizations, and research institutions should work together to develop global conservation policies for soil biodiversity protection (Hendrix & Bohlen, 2002).
- Legislation and Policy Integration: Soil health should be incorporated into national biodiversity conservation strategies, emphasizing sustainable land management and pollution control.
- Public Engagement and Education: Increasing awareness about the ecological importance of earthworms can drive community-led conservation efforts. Initiatives like citizen science projects and educational programs can help monitor earthworm populations globally.

1.15 SYSTEMATICS

Domain: Eukaryota—Woese, Kandler & Wheelis, 1990
Kingdom: Animalia—Haeckel, 1878
Superphylum: Lophotrochozoa—Halanych et al., 1995

Phylum: Annelida—Lamarck, 1802
Class: Clitellata—Michaelsen, 1919
Subclass: Oligochaeta—Grube, 1850
Supraorder: Megadrili—Benham, 1890
Order: Crassiclitellata—Jamieson, 1988
Suborder: Lumbricina—de Blainville, 1830

1. Family: Acanthodrilidae (Claus, 1880a,b), comprising 68 genera, 746 species.
   Distribution: The species has a broad distribution across multiple continents and regions. It is found in South America, South Africa, Madagascar, Australia, New Zealand, New Caledonia, and the Antarctic Islands (Fragoso & Rojas, 2016, 2019; Buckley et al., 2011). In North America, it has been recorded in the USA and Mexico, extending to the Caribbean Islands (Reynolds, 2020; Fragoso & Rojas, 2016). Additionally, its presence is noted in Europe and Northern Asia, as well as Central America and the Caribbean Islands, including Mexico (Fragoso & Rojas, 2016, 2018, 2019; Cervantes et al., 2016). The species also occurs in eastern Africa and Madagascar (Csuzdi et al., 2016; Hong et al,. 2019). Furthermore, it is distributed across the Andean region, Chile, and Argentina (Fragoso & Rojas, 2016, 2018; Cervantes & Fragoso, 2018).
2. Family: Almidae (Duboscq, 1902), comprising 7 genera, 64 species.
   Distribution: The species is widely distributed across Africa, Southeast Asia, South America, and the Caribbean. Within Africa, it has been recorded in various regions, while in South America and the Caribbean, its range extends from Costa Rica to southeastern Brazil, including the Amazonia region and Guarani, Uruguay (Grosso & Brown, 2007).
3. Family: Arecoidae (James et al., 2023), comprising 1 genus, 1 species.
   Distribution: The species is also recorded in Brazil (Misirlioğlu et al., 2023).
4. Family: Benhamiidae (Michaelsen, 1897), comprising 20 genera, 351 species.
   Distribution: The species has a widespread distribution across multiple continents and regions. In Africa, it is found in Saharan, central, and eastern Africa (Gates, 1972; Csuzdi, 2010). It is also recorded in Oceania, including the Pacific islands of Fiji and Samoa (Easton, 1984; James & Davidson, 2012). In the Americas, its presence extends across Central America and northern South America (Csuzdi, 2010), as well as southern and central Brazil (Bartz et al., 2009; Bartz et al., 2014; Santos et al., 2018). Additionally, it has been documented in Europe, specifically in Hungary, Ireland, Finland, Sweden, and Israel (Terhivuo, 1991; Erséus et al., 1994; Rota & Schmidt, 2006; Csuzdi et al., 2008). The species is further distributed across Africa, Southeast Asia, Australia, Oceania, Central America, the Caribbean, and southern North America (Blakemore, 2010).
5. Family: Biwadrilidae (Jamieson, 1971), comprising 1 genus, 1 species.
   Distribution: The species is also recorded in Japan (Brinkhurst, 1981).
6. Family: Criodrilidae (Vejdovský, 1884), comprising 1 genus, 2 species.

Distribution: The species is distributed across northern Europe (Smiljkov et al., 2005; Blakemore, 2008; Atanacković et al., 2013; Valchovski, 2013; Mısırlıoğlu, 2017; Timm, 2020). In North America, it has been recorded in the USA (McKey-Fender & MacNab, 1953), while in South America, its presence is noted in Brazil (Knäpper, 1976; Knäpper & Porto, 1979).

7. Family: Diporodrilidae (Bouché, 1970), comprising 1 genus, 3 species.
   Distributed: The species is also recorded in the Mediterranean islands of Corsica and Sardinia (Marchán et al., 2022).
8. Family: Eudrilidae (Claus, 1880a,b), comprising 45 genera, 305 species.
   Distribution: The species has a broad distribution across multiple continents. In Africa, it has been recorded in various regions (Sims, 1987), including specific locations such as Ivory Coast, Nigeria, Gabon, and central Africa, encompassing Congo, the Democratic Republic of Congo, and Angola (Tondoh & Lavelle, 2005). In Latin America and the Caribbean, its range extends to Brazil (Righi, 1972; Sousa et al., 2020). Additionally, it is found in Madagascar, Sri Lanka, India, the Philippines, Malaysia, Indonesia, Vietnam, and Australia (Blakemore, 2015).
9. Family: Glossoscolecidae (Michaelsen, 1900), comprising 6 genera, 156 species.
   Distribution: The species is widely distributed across Central and South America, ranging from Panama to northern Argentina and Uruguay. It is also recorded in southeastern Brazil (Feijoo & Brown, 2023). In the Caribbean, its presence has been documented in Guadeloupe (James & Gamiette, 2016). Additionally, it is found from Argentina to Venezuela (Fragoso & Brown, 2007a, 2007b) and in northern South America, including southeastern Brazil (Silva et al., 2017).
10. Family: Hormogastridae (Michaelsen, 1900), comprising 9 genera, 37 species.
    Distribution: The genera *Hormogaster* and *Boucheona* have been documented by Marchán et al. (2018). Their distribution extends across Europe and northern Africa, including southern France, the Spanish Iberian Peninsula, Corsica, Sardinia, the Tuscan Archipelago, Italy, and Sicily. In northern Africa, they are recorded in Algeria and Tunisia, while their presence is also noted in Portugal (Marchán et al., 2023b).
11. Family: Kazimierzidae (Nxele et al., 2016), comprising 1 genus, 25 species.
    Distribution: The species is also recorded in the South African Western and Northern Cape provinces (Myers et al., 2000).
12. Family: Komarekionidae (Gates, 1974), comprising 1 genus, 1 species.
    Distribution: The species is distributed in the Mid-Atlantic states of the USA, extending west to southern Illinois (Gates, 1974; Reynolds, 2020).
13. Family: Kynotidae (Jamieson, 1971), comprising 1 genus, 22 species.
    Distribution: The species is also recorded in Madagascar (Razafindrakoto et al., 2017).

14. Family: Lumbricidae (Rafinesque, 1815), comprising 47 genera, 615 species.
    Distribution: The species has an extensive distribution across multiple continents. In North America, it is recorded in eastern North America (Brown et al., 2023) and more broadly across the continent (Fragoso & Rojas, 2014; Reynolds, 2020, 2022). Also reported in the Mid-Atlantic states, extending west to southern Illinois (Gates, 1974; Reynolds, 2020), as well as the North American Arctic on Ellesmere Island (Francis, 1991; Eberle & Greenwood, 2012). In South America, its range includes southern South America (Zicsi, 2007) and other regions (Fragoso & Brown, 2007a,b). The species is documented in Europe, with records from various locations, including the Balkan region (Popović et al., 2022; Trakić et al., 2016), the Apuseni Mountains in the southern Carpathians (Csuzdi et al., 2011), and from southern France to the northern Iberian Peninsula, including Corsica (Marchán et al., 2023a). It is also present in Switzerland, the Czech Republic, Turkey, and Asia's central part (Mısırlıoğlu, 2017; Omodeo, 1952, 1956, 1961, 1988; Bouché, 1972, 1983; Marchán et al. 2020, 2022b; Mršić & Šapkarev, 1988; Mršić, 1991).
    In Africa, the species has been recorded in various regions, including Madagascar (Razafindrakoto et al., 2017), South Africa's Western Cape and Northern Cape provinces (Myers et al., 2000), and other locations (Omodeo, 1952, 1956, 1961, 1988; Bouché, 1972, 1983; Mısırlıoğlu, 2017). Additionally, its distribution extends to Australia, New Zealand, and the Pacific islands (Zicsi, 2007).
15. Family: Lutodrilidae (McMahan, 1976), comprising 1 genus, 1 species.
    Distribution: The species is also recorded in eastern Louisiana, USA (McMahan, 1976; Anderson et al., 2017).
16. Family: Megascolecidae (Rosa, 1891), comprising 85 genera, 2208 species.
    Distribution: The species has a widespread distribution across multiple continents and regions. In Australia and New Zealand, it is recorded in both eastern and southeastern parts (Sims, 1980; Jamieson, 2000). In North America, it is found in the Pacific Northwestern states of the United States, southern Canada, and California (Reynolds, 2020, 2022). Additionally, it has been documented in eastern Louisiana (McMahan, 1976; Anderson et al., 2017). Its presence is noted in multiple locations, including San Francisco, Rio de Janeiro, Hawaii, and Peru (Brown et al., 2006a, 2006b; Chang et al., 2017). It is also recorded more broadly across North America and Asia (Buckley et al., 2011). The species has been documented in the Pacific Ocean region (Blakemore, 2007; James, 2011) and in the USA (Chang et al., 2021).
17. Family: Microchaetidae (Michaelsen, 1900), comprising 3 genera, 81 species.
    Distribution: The species is distributed across Mediterranean southern Africa, including northern South Africa, Lesotho, and Swaziland

(Plisko, 2013). Its range also extends to Zimbabwe, Botswana, and Mozambique (Plisko, 1992, 2013).

18. Family: Moniligastridae (Claus, 1880a,b), comprising 5 genera, 185 species.
    Distribution: The species is widely distributed across Asia, extending from India east to Philippines and north to southern Siberia and Japan (Perel, 1997; Gates, 1972). It is also recorded in Africa (Csuzdi, 2005), Australia, and Southeast Asia. Additionally, its range includes Mexico, Central America, the Caribbean, and several Pacific Islands (Blakemore et al., 2014; Csuzdi et al., 2017).
19. Family: Ocnerodrilidae (Beddard, 1891), comprising 37 genera, 172 species.
    Distribution: Broadly distributed across multiple continents. In Asia, it is found in India, China, and Myanmar (Fragoso & Rojas, 2009a,b). Its range extends across South and Central America, sub-Saharan Africa, and India. In Brazil, it has been documented in various regions (James et al., 2023; Hernández-García et al., 2018), including southern Brazil (Bartz et al., 2013; Bartz et al., 2014; Ferreira et al., 2023; Lima & Rodríguez, 2007). Additionally, the species is recorded across Latin America, Africa, and Asia, with a possible Central Pangean origin linking its presence in South America and Africa (Fragoso & Rojas, 2009a, 2009b).
20. Family: Rhinodrilidae (Benham, 1890), comprising 38 genera, 376 species.
    Distribution: The species is widely distributed across South and Central America. It is recorded in Argentina (Mischis, 2007) and extends northward up to Mexico (Fragoso & Rojas, 2014). In the Caribbean, it is found on various islands (Rodríguez et al., 2007). Its range includes Chile (Zicsi & Csuzdi, 2007) and spans Panama in the north to Peru in the south, reaching as far east as French Guiana (Fragoso & Brown, 2007a, 2007b). Additionally, it is documented in Paraguay (Brown & Fragoso, 2007) and Brazil (Römbke et al., 1999; Zicsi et al., 2001).
21. Family: Sparganophilidae (Michaelsen, 1918), comprising 1 genus, 12 species.
    Distribution: The species has a broad distribution across North America, including the USA and Mexico (Ikeda et al., 2020) and Canada (Reynolds, 2022). It is also recorded in Europe, with occurrences in Germany, France, England, Switzerland, Italy, and Spain (Rota et al., 2018; Bouché & Qiu, 1998). Additionally, it has been documented across North America in earlier studies (Reynolds, 1980, 2008a, 2008b).
22. Family: Tritogeniidae (Plisko, 2013), comprising 2 genera, 39 species.
    Distribution: Southern Africa, Botswana (Nxele et al., 2018).
23. Family: Tumakidae (Righi, 1995), comprising 1 genus, 3 species.
    Distribution: The species is distributed in central Colombia and the Caribbean region, particularly along the Magdalena River valley (Righi, 1995; Celis & Rangel-Ch., 2015).

1.16 DISCUSSION

The global diversity and distribution of earthworms, encompassing 5406 species within 382 genera and 23 families, highlight their critical ecological roles and adaptive success in various environments (Mısırlıoğlu et al., 2023). Families such as Acanthodrilidae are widely distributed across South America, South Africa, Madagascar, Australia, New Zealand, and North America, while Almidae inhabit Africa, Southeast Asia, and South America. The monotypic Arecoidae is endemic to Brazil, whereas Benhamiidae spans Saharan Africa, Oceania, and Central and South America. The Biwadrilidae, Criodrilidae, and Diporodrilidae are more localized, found in Japan, northern Europe, and the Mediterranean islands, respectively. The Eudrilidae family is prevalent in Africa, Latin America, the Caribbean, and parts of Asia and Australia, whereas Glossoscolecidae are primarily found in Central and South America. Families such as Hormogastridae are limited to Europe and northern Africa, while Kazimierzidae are restricted to South Africa. The Komarekionidae occur in the USA, and Kynotidae are found exclusively in Madagascar. Widely distributed families like Lumbricidae span Europe, North America, Asia, and Australia, while Lutodrilidae is restricted to Louisiana, USA. The highly diverse Megascolecidae occur across Australia, Asia, and North America, with Microchaetidae centered in southern Africa. The Moniligastridae family extends from India to Southeast Asia, while Ocnerodrilidae spans South America, Africa, and parts of Asia. The Rhinodrilidae are found from Argentina to Mexico, while Sparganophilidae span North America and parts of Europe. Tritogeniidae occur in southern Africa, and Tumakidae are confined to Colombia and the Caribbean. These families collectively underline the remarkable adaptive capacity of earthworms to diverse ecological conditions, from the tropics to temperate zones. Their contributions to soil health, nutrient cycling, and agricultural productivity, coupled with human-facilitated introductions of species like *Eudrilus eugeniae*, underscore their ecological importance and the need for sustainable management.

In the next chapters we discuss the diversity of earthworms in India. We also discuss ecology, anatomy, the lifecycle gut microbiome of earthworms, and their economic importance.

REFERENCES

Aira, M, Sampedro, L, Monroy, F, & Dominguez, J. (2008). Detritivorous earthworms directly modify the structure, thus altering the functioning of a microdecomposer food web. *Soil Biology and Biochemistry*, 40, 2511–2516.

Anderson, F. E., Williams, B. W., Horn, K. M., Erséus, C., Halanych, K. M., Santos, S. R., & James, S. W. (2017). Phylogenomic analyses of Crassiclitellata support major Northern and Southern Hemisphere clades and a Pangaean origin for earthworms. *BMC Evolutionary Biology*, 17(1), 123.

Atanacković, A. D., Šporka, F., Csányi, B., Vasiljević, B. M., Tomović, J. M., & Paunović, M. M. (2013). Oligochaeta of the Danube River—a faunistical review. *Biologia*, 68(2), 269–277.

Ayuke, F. O., Brussaard, L., Vanlauwe, B., Six, J., Lelei, D. K., Kibunja, C. N., & Pulleman, M. M. (2011). Soil fertility management: Impacts on soil macrofauna, soil aggregation, and soil organic matter allocation. *Applied Soil Ecology*, 48(1), 53–62.

Baker, G. H., Brown, G., Butt, K., Curry, J. P., & Scullion, J. (2006). Introduced earthworms in agricultural and reclaimed land—Their ecology and influences on soil properties, plant production and other soil biota. *Biological Invasions*, 8, 1301–1316.

Baker, G. H., Carter, P. J., & Barrett, V. J. (1999). Survival and biomass of exotic earthworms, Aporrectodea spp. (Lumbricidae), when introduced to pastures in south-eastern Australia. *Australian Journal of Agricultural Research*, 50(7). (In press.)

Bartz, M. L. C., Brown, G. G., da Rosa, M. G., Filho, O. K., James, S. W., Decaëns, T., & Baretta, D. (2014). Earthworm richness in land-use systems in Santa Catarina, Brazil. *Applied Soil Ecology*, 83, 59–70.

Bartz, M. L. C., Brown, G. G., & James, S. W. (2013). Earthworms as soil quality indicators in Brazilian no-tillage systems. *Applied Soil Ecology*, 64, 1–8. https://doi.org/10.1016/j.apsoil.2012.10.007

Bartz, M. L. C., Brown, G. G., Pasini, A., de Oliveira Fernandes, J., Curmi, P., Dorioz, J., & Ralisch, R. (2009). Earthworm communities in organic and conventional coffee cultivation. *Pesquisa Agropecuária Brasileira*, 44, 928–933.

Beddard, F. E. (1891). On the anatomy of Ocnerodrilus. *Transactions of the Royal Society of Edinburgh*, 36, 563–583 + Plate XXXVI.

Benham, W. B. (1890). An attempt to classify earthworms. *Quarterly Journal of the Microscopical Society*, 31, 201–315.

Blakemore, R. J. (2006). *Cosmopolitan Earthworms—an Eco-Taxonomic Guide to the Peregrine Species of the World*. VermEcology.

Blakemore, R. J. (2007). Origin and means of dispersal of cosmopolitan Pontodrilus litoralis (Oligochaeta: Megascolecidae). *European Journal of Soil Biology*, 43, S3–S8.

Blakemore, R. J. (2008). Review of Criodrilidae (Annelida: Oligochaeta) including Biwadrilus from Japan. *Opuscula Zoologica Budapest*, 37, 11–22.

Blakemore, R. J. (2010). *Cosmopolitan Earthworms—an Eco-Taxonomic Guide to the Peregrine Species of the World* (4th ed.). Yokohama: VermEcology.

Blakemore, R. J. (2012). *Cosmopolitan Earthworms—An Eco-Taxonomic Guide to the Peregrine Species of the World*. VermEcology.

Blakemore, R. J. (2015). Eco-taxonomic profile of an iconic vermicomposter—the 'African Night-crawler' earthworm Eudrilus eugeniae (Kinberg, 1867). *African Invertebrates*, 56(3), 527–548.

Blakemore, R. J. (2018). Global distribution of earthworm diversity. *Science*, 365(6457), 144–145. https://www.science.org/doi/10.1126/science.aax4851

Blakemore, R. J., Lee, S., & Seo, H.-Y. (2014). Reports of *Drawida* (Oligochaeta: Moniligastridae) from Far East Asia. *Journal of Species Research*, 3(1), 1–9. https://doi.org/10.12651/JSR.2014.3.1.1

Blanchart, E., Lavelle, P., Braudeau, E., Le Bissonnais, Y., & Valentin, C. (2004). Regulation of soil structure by geophagous earthworm activities in humid savannas of Côte d'Ivoire. *Soil Biology and Biochemistry*, 36(2), 231–245.

Blouin, M., Hodson, M. E., Delgado, E. A., Baker, G., Brussaard, L., Butt, K. R., Dai, J., Dendooven, L., Peres, G., Tondoh, J. E., Cluzeau, D., & Brun, J. J. (2013). A review of earthworm impact on soil function and ecosystem services. *European Journal of Soil Science*, 64(2), 161–182. https://doi.org/10.1111/ejss.12025

Bohlen, P. J., Groffman, P. M., Fahey, T. J., Fisk, M. C., Suarez, E., Pelletier, D. M., & Fahey, R. T. (2004a). Ecosystem consequences of exotic earthworm invasion of north temperate forests. *Ecosystems*, 7(1), 1–12.

Bohlen, P. J., Scheu, S., Hale, C. M., McLean, M. A., Migge, S., Groffman, P. M., & Parkinson, D. (2004b). Non-native invasive earthworms as agents of change in northern temperate forests. *Frontiers in Ecology and the Environment*, 2(8), 427–435.

Bouché, M. B. (1970). Remarques sur quelques Lumbricina de France et consequences de la decouverte des nouveaux taxons Vignysinae (Subfam. nov.) et Diporodrilidae (Fam. nov.). *Pedobiologia*, 10(1), 246–256.

Bouché, M. B. (1972). *Lombriciens de France: Écologie et Systématique* (p. 67). Institut National de la Recherche Agronomique, Articles de Zoologie-Écologie animale (Numéro hors-série).

Bouché, M. B. (1977). Strategies lombriciennes. In U. Lohm, & T. Persson (Eds.), *Soil Organisms as Components of Ecosystems.* Proceedings of the VI International Soil Zoology Colloquium, Uppsala, Sweden, 1974. *Ecological Bulletins (Stockholm)*, 25, 122–132.

Bouché, M. B. (1983). The establishment of earthworm communities. In J. E. Satchell (Ed.), *Earthworm Ecology from Darwin to Vermiculture* (pp. 431–448). London, UK: Chapman & Hall.

Bouché, M. B., & Qiu, J.-P. (1998). Un nouveau Sparganophilus (Annelida: Oligochaeta) d'Europe, avec considérations paléogéographiques sur les Lumbricina. *Documents Pédozoologiques et Intégrologiques*, 4, 178–180.

Brinkhurst, R. O. (1981). A contribution towards understanding the earthworm fauna of Japan. *Proceedings of the Japanese Society of Systematic Zoology*, 20, 93–109.

Brown, G. G., Barois, I., & Lavelle, P. (2006a). Regulation of soil organic matter dynamics and microbial activity in the drilosphere and the role of interactions with other edaphic functional domains. *European Journal of Soil Biology*, 42, S205–S208. https://doi.org/10.1016/j.ejsobi.2006.10.002

Brown, G. G., & Fragoso, C. (2007). Listado de las especies de lombrices de tierra de América Central Continental (Guatemala, Belice, Honduras, El Salvador, Nicaragua, Costa Rica, Panamá), las Guyanas (Surinam, Guyanne Française, Guyana), Venezuela y Paraguay. In G. G. Brown, & C. Fragoso (Eds.), *Minhocas na América Latina: Biodiversidade e Ecologia* (pp. 422–452). Londrina: Embrapa Soja.

Brown, G. G., James, S. W., Csuzdi, C., Lapied, E., Decaëns, T., Reynolds, J. W., Mısırlıoğlu, M., Stovanic, M., Trakić, T., Sekulić, J., Phillips, H., & Cameron, E. (2023). *A Checklist of Megadrile Earthworm (Annelida: Clitellata) Species and Subspecies of the World.* Available from: Zenodo.

Brown, G. G., James, S. W., Pasini, A., Nunes, D. H., Benito, N. P., Martins, P. T., & Sautter, K. D. (2006b). Exotic, peregrine, and invasive earthworms in Brazil: Diversity, distribution, and effects on soils and plants. *Caribbean Journal of Science*, 42, 339–358.

Buckley, T. R., James, S., Allwood, J., Bartlam, S., Howitt, R., & Prada, D. (2011). Phylogenetic analysis of New Zealand earthworms (Oligochaeta: Megascolecidae) reveals ancient clades and cryptic taxonomic diversity. *Molecular Phylogenetics and Evolution*, 58, 85–96.

Callaham, M. A., Jr., Hendrix, P. F., & Phillips, R. J. (2002). Occurrence of an exotic earthworm (*Amynthas agrestis*) in undisturbed soils of the Southern Appalachian Mountains, USA. *Pedobiologia*, 46(5), 407–413.

Celis, L., & Rangel-Ch, O. (2015). Dos especies nuevas (Oligochaeta: Tumakidae) de la región Caribe de Colombia. *Papéis Avulsos de Zoologia*, 55(30), 405–414.

Cervantes, G., & Fragoso, C. (2018). Protozapotecia acaxetlensis and Protozapotecia oyametlensis, two new earthworm species (Crassiclitellata: Acanthodrilidae) from Mexican central mountains. *Zootaxa*, 4496, 431–439.

Cervantes, G., Fragoso, F., Monteros, A. E., Sánchez-Ramos, G., Lara-Villalón, M., Yañez-Pacheco, M. J., Lázaro-Castellanos, J. O., & James, S. W. (2016). New species of the earthworm genus Zapatadrilus (Clitellata, Acanthodrilidae) from Northern Mexico. *Zootaxa*, 4189(2), 311–326.

Chang, C.-H., Bartz, M. L. C., Brown, G., Callaham, M. A., Jr., Cameron, E. K., Dávalos, A., Dobson, A., Görres, J. H., Herrick, B. M., Ikeda, H., James, S. W., Johnston, M. R., McCay,

T. S., McHugh, D., Minamiya, Y., Nouri-Aiin, M., Novo, M., OrtizPachar, J., Pinder, R. A., Ransom, T., Richardson, J. B., Snyder, B. A., & Szlavecz, K. (2021). The second wave of earthworm invasions in North America: Biology, environmental impacts, management and control of invasive jumping worms. *Biological Invasions*, 23, 3291–3322.

Chang, C. H., & Chen, J. H. (2005) Three new species of octothecate pheretimoid earthworms from Taiwan, with discussion on the biogeography of related species. *Journal of Natural History*, 39, 1469–1482.

Chang, C.-H., Snyder, B., & Szlavecz, K. (2017). Asian pheretimoid earthworms in North America north of Mexico: An illustrated key to the genera Amynthas, Metaphire, Pithemera, and Polypheretima (Clitellata: Megascolecidae). *Zootaxa*, 4179(3), 495–529.

Claus, C. (1880a). *Grundzüge der Zoologie* (5th edition, 822 pp.). Marburg: N.G. Elwertsche Universitäts-Buchhandlung.

Claus, C. (1880b). Über die systematische Stellung der Moniligastriden. *Zoologischer Anzeiger*, 3, 449–452.

Csuzdi, Cs. (2005). Earthworms (Annelida: Oligochaeta) of Sao Tome. *Journal of Natural History*, 39(33), 3039–3058.

Csuzdi, Cs. (2010). *A Monograph of the Paleotropical Benhamiinae Earthworms (Annelida: Oligochaeta, Acanthodrilidae)* (p. 348). Budapest: Hungarian Natural History Museum.

Csuzdi, Cs. (2012). Earthworm species, a searchable database. *Opuscula Zoologica*, 43(1), 97–99. https://doi.org/10.5281/zenodo.1045525

Csuzdi, Cs., Pavlícek, T., & Nevo, E. (2008). Is Dichogaster bolaui (Michaelsen, 1891) the first domicole earthworm species? *European Journal of Soil Biology*, 44, 198–201.

Csuzdi, Cs., Pearlson, O., & Pavlícek, T. (2017). New Acanthodrilus species from New Caledonia (Clitellata, Megadrili, Acanthodrilidae). *Journal of Natural History*, 51, 1899–1912.

Csuzdi, Cs., Pop, V. V., & Pop, A. A. (2011). The earthworm fauna of the Carpathian Basin with new records and description of three new species (Oligochaeta: Lumbricidae). *Zoologischer Anzeiger*, 250, 2–18.

Csuzdi, Cs., Razafindrakoto, M., & Hong, Y. (2016). The second species of the endemic Malagasy earthworm genus Howascolex Michaelsen, 1901; Howascolex farafangana sp. n. (Clitellata, Megadrili). *African Invertebrates*, 57(2), 83–91.

Curry, J. P., & Schmidt, O. (2007). The feeding ecology of earthworms—a review. *Pedobiologia*, 50(6), 463–477.

Decaëns, T., Jiménez, J. J., Barros, E., Chauvel, A., Blanchart, E., Fragoso, C., & Lavelle, P. (2006). Soil macrofaunal communities in permanent pastures derived from tropical forest or savanna. *Agriculture, Ecosystems & Environment*, 114(2–4), 405–415. https://doi.org/10.1016/j.agee.2005.11.022

Decaëns, T., Lavelle, P., Jiménez, J. J., Escobar, G., & Rippstein, G. (1999). Impact of land management on soil macrofauna in the Oriental Llanos of Colombia. *European Journal of Soil Biology*, 35(4), 157–168.

DriloBase. (2022). The world earthworm database. *Drilobase Project*. Available from: http://drilobase.org (Accessed 7 November 2022).

Duboscq, O. (1902). Études sur les Oligochètes. *Archives de Zoologie Expérimentale et Générale*, 40, 273–322.

Dyne, G. R. (1979). A new species of bioluminescent earthworm from North Queensland belonging to the subgenus Diplotrema (Oligochaeta: Megascolecidae). *Memoirs of the Queensland Museum*, 19(3), 373–379.

Dyne, G. R. (1997). Large earthworms (Megascolecidae) from Australia: A taxonomic revision. *Records of the Australian Museum*, 49(1), 1–43.

Easton, E. G. (1984). Earthworms (Oligochaeta) from islands of the South-Western Pacific, and a note on two species from Papua New Guinea. *New Zealand Journal of Zoology*, 11, 111–128.

Eberle, J. J., & Greenwood, D. R. (2012). Life at the top of the greenhouse Eocene World—a review of the Eocene flora and vertebrate fauna from Canada's high Arctic. *Geological Society of America Bulletin*, 124, 3–23.

Edwards, C. A., & Arancon, N. Q. (2004). The use of earthworms in the breakdown of organic wastes to produce vermicomposts and animal feed protein. *Earthworm Ecology*, 2, 345–438.

Edwards, C. A., & Bohlen, P. J. (1996a). *Biology and Ecology of Earthworms* (3rd ed.). Springer Science & Business Media.

Edwards, C. A., & Bohlen, P. J. (1996b). *Biology and Ecology of Earthworms* (3rd ed.). Chapman & Hall.

Eggleton, P., Inward, K., Smith, J. J., Jones, D. T., & Sherlock, E. (2009). A six-year study of earthworm (Lumbricidae) vertical distribution in grassland in Southern England shows seasonal and spatial stability. *Soil Biology and Biochemistry*, 41(12), 261–271. https://doi.org/10.1016/j.soilbio.2009.07.048

Eisenhauer, N., & Straube, D. (2021). Climate change effects on earthworms: A review. *Soil Organisms*, 81(3), 585–608.

Eisenhauer, N., Straube, D., & Scheu, S. (2014). Climate change effects on earthworms: A review. *Soil Organisms*, 81(3), 585–608.

Erséus, C., Omodeo, P., & Rota, E. (1994). The first records of the allochthonous species Dichogaster saliens and D. bolaui (Oligochaeta: Octochaetidae) from Sweden. *Megadrilogica*, 6(2), 17–20.

Feijoo, A., & Brown, G. G. (2023). Three new Glossoscolex (Annelida: Crassiclitellata: Glossoscolecidae) in the truncatus group from the Brazilian Atlantic Forest. *Zootaxa*, 5225(1), 220–234.

Ferlian, O., Eisenhauer, N., Aguirrebengoa, M., Camara, M., Ramirez-Rojas, I., Santos, F., & Tanalgo, K. (2020). Invasive earthworms erode soil biodiversity: A meta-analysis. *Journal of Animal Ecology*, 89(3), 537–547. https://doi.org/10.1111/1365-2656.13104

Ferreira, T., James, S. W., Bartz, M. L. C., Lima, A. C. R., Dudas, R., & Brown, G. G. (2023). Distribution and diversity of earthworms in different land use systems in Rio Grande do Sul, Brazil. *Zootaxa*, 5225(1), 389–398.

Fierer, N., M. A. Bradford, and R. B. Jackson. (2009). Global patterns in belowground communities. *Ecology Letters*, 12, 637–649.

Fragoso, C., & Brown, G. G. (2007a). Ecología y taxonomía de las lombrices de tierra en latinoamérica: El Primer Encuentro LatinoAmericano de Ecología y Taxonomía de Oligoquetos (ELAETAO1). In G. G. Brown, & C. Fragoso (Eds.), *Minhocas na América Latina: Biodiversidade e Ecologia* (pp. 33–75). Londrina: Embrapa Soja.

Fragoso, C., & Brown, G. G. (2007b). Ecological and taxonomic diversity of earthworms in Brazilian ecosystems. In A. M. R. Pinto, E. G. de A. Coutinho, & W. M. G. Rodrigues (Eds.), *Minhocas na América Latina: Biodiversidade e Ecologia* (pp. 67–109). Embrapa Solos.

Fragoso, C., Brown, G. G., & Lavelle, P. (1999). Soil biodiversity and sustainable agriculture: Earthworm diversity and functioning in managed ecosystems. In D. Pimentel (Ed.), *Agroecosystem Sustainability: Developing Practical Strategies* (pp. 263–289). CRC Press.

Fragoso, C., Brown, G. G., Patrón, J. C., Blanchart, E., Lavelle, P., Pashanasi, B., & Kumar, T. (1997a). Agricultural intensification, soil biodiversity and agroecosystem function in the tropics: The role of earthworms. *Applied Soil Ecology*, 6(1), 3–16.

Fragoso, C., Moreno, A., Kanyonyo, J., Blanchart, E., Rodriguez, C., & Senapati, B. (1997b). A survey of tropical earthworms: Taxonomy, bio-geography and environmental plasticity. In P. Lavelle (Ed.), *The Potential for Earthworm Management in Tropical Agroecosystems*. CAB International. (In press.)

Fragoso, C., & Rojas, P. (2009a). A new ocnerodrilid earthworm genus from Southeastern Mexico (Annelida: Oligochaeta), with a key for the genera of Ocnerodrilidae. *Megadrilogica*, 13(9), 141–152.

Fragoso, C., & Rojas, P. (2009b). Global distribution of earthworm diversity. *Science*, 365(6457), 144–145. https://doi.org/10.1126/science.aax4851

Fragoso, C., & Rojas, P. (2014). Biodiversidad de lombrices de tierra (Annelida: Oligochaeta: Crassiclitellata) en México. *Revista Mexicana de Biodiversidad, Suplemento*, 85, S197–S207.

Fragoso, C., & Rojas, P. (2016). Lavellodrilus notosetosus sp. nov. (Annelida, Crassiclitellata, Acanthodrilidae): A new Mexican earthworm with uncommon characters, revealed by a preliminary revision of subfamily Acanthodrilinae. *Zootaxa*, 4154, 101–138.

Fragoso, C., & Rojas, P. (2018). New Diplotrema and Lavellodrilus earthworm species from Southern Mexico (Annelida, Crassiclitellata, Acanthodrilidae). *Zootaxa*, 4496(1), 414–430.

Fragoso, C., & Rojas, P. (2019). More new Diplotrema earthworm species from Southern Mexico (Annelida, Crassiclitellata, Acanthodrilidae, Acanthodrilinae). *Zootaxa*, 4688(4), 483–502.

Francis, J. E. (1991). The dynamics of polar fossil forests: Tertiary fossil forests of Axel Heiberg Island, Canadian Arctic Archipelago. *Geological Survey of Canada, Bulletin*, 403, 29–38.

Frelich, L. E., Hale, C. M., Scheu, S., Holdsworth, A. R., Heneghan, L., Bohlen, P. J., & Reich, P. B. (2006). Earthworm invasion into previously earthworm-free temperate and boreal forests. *Biological Invasions*, 8(6), 1235–1245.

Gates, G. E. (1972). Burmese earthworms: An introduction to the systematics and biology of Megadrile Oligochaetes with special reference to Southeast Asia. *Transactions of the American Philosophical Society*, 62, 1–326.

Gates, G. E. (1974). On a new species of earthworm in a southern portion of the United States. *Bulletin of Tall Timbers Research Station*, 15, 1–13.

Görres, J. H., Bellitürk, K., & Melnichuk, R. D. (2019). Abundance and ecological impact of invasive *Amynthas* earthworms in temperate forests. *Biological Invasions*, 21(6), 2081–2095.

Grosso, E., & Brown, G. G. (2007). Biodiversidad y ecología de las lombrices de tierra en el Uruguay. In G. G. Brown, & C. Fragoso (Eds.), *Minhocas na América Latina: Biodiversidade e Ecologia* (pp. 281–296). Londrina: Embrapa Soja.

Gundale, M. J., Jolly, W. M., & Deluca, T. H. (2005). Susceptibility of a northern hardwood forest to exotic earthworm invasion. *Conservation Biology*, 19(4), 1075–1083. https://doi.org/10.1111/j.1523-1739.2005.00102.x

Hendrix, P. F., & Bohlen, P. J. (2002). Exotic earthworm invasions in North America: Ecological and policy implications. *BioScience*, 52(9), 801–811.

Hendrix, P. F., Callaham, M. A., Jr., Drake, J. M., Huang, C. Y., James, S. W., Snyder, B. A., & Zhang, W. (2008). Pandora's box contained bait: The global problem of introduced earthworms. *Annual Review of Ecology, Evolution, and Systematics*, 39, 593–613. https://doi.org/10.1146/annurev.ecolsys.39.110707.173426

Hernández-García, L. M., Burgos-Guerrero, J. E., Rousseau, G. X., & James, S. W. (2018). Brasilisia n. gen. and Arraia n. gen., two new genera of Ocnerodrilidae (Annelida, Clitellata, Oligochaeta) from Eastern Amazonia, Brazil. *Zootaxa*, 4496, 471–480.

Hong, Y., Razafindrakoto, M., Blanchart, E., Szederjesi, T., & Csuzdi, Cs. (2019). New acanthodrilid species from Madagascar (Clitellata, Acanthodrilidae). *Acta Zoologica Academiae Scientiarum Hungaricae*, 65(3), 215–233.

Ikeda, H., Callaham, M. A., Jr., Shefferson, R. P., Wenk, E. S., & Fragoso, C. (2020). A comparison of latitudinal species diversity patterns between riverine and terrestrial earthworms from the North American temperate zone. *Journal of Biogeography*, 47, 1373–1382.

Integrated Taxonomic Information System [ITIS]. (2022). *On-Line Database.* Available from: https://www.itis.gov/ (Accessed 7 November 2022).

James, S. W. (2011). Earthworms. In D. Simberlof, & M. Rejmánek (Eds.), *Encyclopaedia of Biological Invasions* (pp. 177–183). Berkeley: University of California Press.

James, S. W., Bartz, M. L. C., & Brown, G. G. (2023). New Ocnerodrilidae genera, species and records from Brazil. *Zootaxa*, 5255(1), 235–269.

James, S. W., & Davidson, S. K. (2012). Molecular phylogeny of earthworms (Annelida: Crassiclitellata) based on 28S, 18S and 16S gene sequences. *Invertebrate Systematics*, 26(2), 213–229.

James, S. W., & Gamiette, F. (2016). New species of Dichogaster Beddard, 1888 (Clitellata: Benhamiidae) with additional records of earthworms from Guadeloupe (French West Indies). *Zootaxa*, 4178(3), 391–408.

Jamieson, B. G. M. (1971). Glossoscolecidae. In R. O. Brinkhurst, & B. G. M. Jamieson (Eds.), *The Aquatic Oligochaeta of the World* (pp. 147–199). Oliver & Boyd.

Jamieson, B. G. M. (2000). *The Native Earthworms of Australia (Megascolecidae Megascolecinae).* Science Publishers, Enfield, USA (CD-ROM).

Jamieson, B. G. M., Tillier, S., Tillier, A., Justine, J. L., Ling, E., James, S., McDonald, K., & Hugall, A. F. (2002). Phylogeny of the Megascolecidae and Crassiclitellata (Annelida, Oligochaeta): Combined versus partitioned analysis using nuclear (28S) and mitochondrial (12S, 16S) rDNA. *Zoosystema*, 24, 707–734.

Julka, J. M. (1993). *Earthworm Resources of India and Their Utilization in Vermiculture.* Zoological Survey of India.

Julka, J. M., & Paliwal, R. (1993). Distribution of earthworms in different agro-climatic regions of India. In G. K. Veeresh, D. Rajagopal, & C. A. Viraktamath (Eds.), *Advances in Management and Conservation of Soil Fauna* (pp. 201–206). Oxford & IBH Publishing Co. Pvt. Ltd.

Knäpper, C. F. U. (1976). Preliminar considerations on the occurrence of Oligochaetes in the Estuary of the Guaíba RS. *Estudos Leopoldenses*, 38, 39–41.

Knäpper, C. F. U., & Porto, R. P. (1979). Ocorrência de Oligoquetas nos solos do Rio Grande do Sul. *Acta Biologica Leopoldensia*, 1, 137–166.

Lavelle, P., Barios, I., Blanchart, E., Brown, G., Brussaard, L., Fragoso, C., Jimenez, J. J., Kajondo, K. K., Angeles Martinez, M. D. L., Moreno, A., Poshanasi, B., Senapati, B. K., & Villenave, C. (1998). Earthworms as a resource in tropical agroecosystems. *Nature Resoure*, 34, 26–41.

Lavelle, P., Decaëns, T., Aubert, M., Barot, S., Blouin, M., Bureau, F., Margerie, P., Mora, P., & Rossi, J. P. (2006). Soil invertebrates and ecosystem services. *European Journal of Soil Biology*, 42, S3–S15. https://doi.org/10.1016/j.ejsobi.2006.10.002

Lavelle, P., & Lapied, E. (2003). Endangered earthworms of Amazonia: An homage to Gilberto Righi. *Pedobiologia*, 47(5–6), 419–427.

Lee, K. E. (1985). *Earthworms: Their Ecology and Relationships with Soils and Land Use.* Academic Press.

Lee, K. E. (1959). The earthworm fauna of New Zealand. *New Zealand Department of Scientific and Industrial Research Bulletin 130*, 1–486.

Lima, A. C. R., & Rodríguez, C. (2007). Earthworm diversity from Rio Grande Do Sul, Brazil, with a new native criodrilid genus and species (Oligochaeta: Criodrilidae). *Megadrilogica*, 11(2), 9–18.

Lubbers, I. M., van Groenigen, K. J., Fonte, S. J., Six, J., Brussaard, L., & van Groenigen, J. W. (2017). Greenhouse-gas emissions from soils increased by earthworms. *Nature Climate Change*, 7(1), 1–6.

Marchán, D. F., Domínguez, J., Hedde, M., & Decaëns, T. (2023a). *The Cradle of Giants: Insight into the Origin of Scherotheca (Lumbricidae, Crassiclitellata) with Description of Eight New Species from Corsica, France.* Zoosystema, submitted.

Marchán, D. F., Fernández, R., de Sosa, I., Sánchez, N., Díaz Cosín, D. J., & Novo, M. (2018). Integrative systematic revision of a Mediterranean earthworm family: Hormogastridae (Annelida, Oligochaeta). *Invertebrate Systematics*, 32, 652–671.

Marchán, D. F., Gérard, S., Hedde, M., Rougerie, R., & Decaëns, T. (2022). An updated checklist and a DNA barcode library for the earthworms (Crassiclitellata, Oligochaeta) of Corsica, France. *Zoosystema*, 44(17), 439–461.

Marchán, D. F., Domínguez, J., Hedde, M., & Decaëns, T. (2023a). The cradle of giants: insights into the origin of Scherotheca Bouché, 1972 (Lumbricidae, Crassiclitellata) with the descriptions of eight new species from Corsica, France. *Zoologi-ska systema*, 45(3), 107–128.

Marchán, D. F., Novo, M., Domínguez, J., Sánchez, N., & Decaëns, T. (2023b). Contribution to the knowledge of the Genus Boucheona (Oligochaeta, Hormogastridae) in France with description of two new species. *Zootaxa*, 5255(1), 68–81.

Marhan, S., Scheu, S., & Kandeler, E. (2015). Climate change effects on earthworm communities and their role in decomposition processes. *Biology and Fertility of Soils*, 51(7), 739–747. https://doi.org/10.1007/s00374-015-1017-z

McKey-Fender, D., & MacNab, J. A. (1953). The aquatic earthworm Criodrilus lacuum Hoffmeister in North America (Oligochaeta Glossoscolecidae). *Wasmann Journal of Biology*, 113, 373–378.

McMahan, M. L. (1976). Preliminary notes on a new megadrile species, genus, and family from the southeastern United States. *Megadrilogica*, 2(11), 6–8.

Michaelsen, W. (1897). Zur Kenntnis der Oligochätenfauna von Ceylon. *Zoologische Jahrbücher. Abteilung für Systematik, Geographie und Biologie der Tiere,* 10, 1–110.

Michaelsen, W. (1900). *Das Tierreich: Oligochaeta*. Berlin: R. Friedländer & Sohn.

Michaelsen, W. (1918). Die Lumbriciden mit besonderer Berücksichtigung der bisher als Familie Glossoscolecidae zusammengefaßten Unterfamilien. *Zoologische Jahrbücher. Abteilung für Systematik, Geographie und Biologie der Tiere,* 41, 1–398.

Mischis, C. (2007). Catálogo de las lombrices de tierra de la Argentina (Annelida, Oligochaeta). In G. G. Brown, & C. Fragoso (Eds.), *Minhocas na América Latina: Biodiversidade e Ecologia* (pp. 247–252). Londrina: Embrapa.

Mısırlıoğlu, M. (2017). *Topraksolucanları, Biyolojileri, Ekolojileri, Zirai Yönleri, Türkiye Türleri ve Türlerin Taksonomik Özellikleri* (2nd ed., p. 168). Ankara: Nobel Yayınları.

Mısırlıoğlu, M., Reynolds, J. W., Stojanović, M., Trakić, T., Sekulić, J., James, S. W., Csuzdi, C., Decaëns, T., Lapied, E., Phillips, H. R. P., Cameron, E. K., & Brown, G. G. (2023). Earthworms (Clitellata, Megadrili) of the world: An updated checklist of valid species and families, with notes on their distribution. *Zootaxa*, 5255(1), 1–152.

Mršić, N. (1991). Monograph on earthworms (Lumbricidae) of the Balkans I, II. *Slovenska Akademija Znanosti in Umetnosti, Razred za Naravoslovne Vede, Opera, Ljubljana*, 31, 1–757.

Mršić, N., & Šapkarev, J. (1988). Revision of the genus Allolobophora Eisen, 1874 (sensu Pop 1941) (Lumbricidae, Oligochaeta). *Acta Musei Macedonici Scientiarum Naturalium*, 19, 1–38.

Myers, N., Mittermeier, R. A., Mittermeier, C. G., Fonseca, G. A. B., & Kent, J. (2000). Biodiversity hotspots for conservation priorities. *Nature*, 403, 853–858.

Nahmani, J., Hodson, M. E., & Black, S. (2007). Effects of metals on life cycle parameters of the earthworm *Eisenia fetida* exposed to field-contaminated soils. *Environmental Pollution*, 149(1), 44–58.

Nxele, T. C., Plisko, J. D., Mwabvu, T., & Zishiri, T. O. (2016). A new family Kazimierzidae for the genus Kazimierzus, earlier recorded to the composite Microchaetidae (Annelida, Oligochaeta). *African Invertebrates*, 57(2), 111–117.

Nxele, T. C., Plisko, J. D., & Ramongalo, B. K. (2018). The first proandric species of Tritogenia Kinberg, 1867 from Botswana, Tritogenia talana sp.n. (Clitellata, Crassiclitellata, Tritogeniidae). *Zootaxa*, 4496, 214–217.

Omodeo, P. (1952). Particolarita della Zoogeografia dei lombrichi. *Bolletino di Zoologia*, 19, 349–369.

Omodeo, P. (1956). Contributo alla revisione dei Lumbricidae. *Archivio Zoologico Italiano*, 41, 129–212.

Omodeo, P. (1961). Le peuplement des Iles Méditerranées et le problème de l'insularité. *Colloques Internationaux du Centre National de la Recherche Scientifique*, 94, 128–133.

Omodeo, P. (1988). The genus Eophila (Lumbricidae, Oligochaeta). *Bolletino di Zoologia*, 55, 73–88.

Omodeo, P. (2000). Evolution and biogeography of megadriles (Annelida, Clitellata). *Italian Journal of Zoology*, 67, 179–201.

Pelosi, C., Barot, S., Capowiez, Y., Hedde, M., & Vandenbulcke, F. (2013). Pesticides and earthworms: A review. *Agronomy for Sustainable Development*, 34(1), 199–228.

Perel, T. S. (1997). *The Earthworms of the Fauna of Russia: Cadaster and Key* (101 pp.). Academy of Sciences of the USSR. Moscow: Nauka. [In Russian.]

Phillips, H. R. P., Cameron, E. K., Ferlian, O., Türke, M., Winter, M., & Eisenhauer, N. (2019). Global distribution of earthworm diversity. *Science*, 365(6457), 144–145. https://doi.org/10.1126/science.aax4851

Plisko, J. D. (1992). The Microchaetidae of Natal, with description of new species of Microchaetus Rapp and Tritogenia Kinberg, and the new genus Proandricus (Oligochaeta). *Annals of the Natal Museum*, 33, 337–378.

Plisko, J. D. (2010). Megadrile earthworms (Oligochaeta: Megascolecidae) of Malawi: A new species and new records. *African Invertebrates*, 51(2), 321–328.

Plisko, J. D. (2013). A new family Tritogeniidae for the genera Tritogenia and Michalakus, earlier accredited to the composite Microchaetidae (Annelida: Oligochaeta). *African Invertebrates*, 54(1), 69–92.

Popović, F., Stojanović, M., Trakić, T., & Sekulić, J. (2022). Earthworms (Annelida: Oligochaeta, Lumbricidae) of the Kopaonik National Park (Serbia). *Kragujevac Journal of Science*, 44, 219–230. https://doi.org/10.46793/kjs44.219P

Rafinesque, C. S. (1815). *Analyse de la nature ou Tableau de l'univers et des corps organisés* (224 pp.). Published by the author, Palermo. https://doi.org/10.5962/bhl.title.106607

Razafindrakoto, M., Csuzdi, C., & Blanchart, E. (2017). New earthworms from Madagascar with key to the Kynotus species (Oligochaeta: Kynotidae). *Zoologischer Anzeiger*, 266, 1–14. https://doi.org/10.1016/j.jcz.2016.08.001

Reynolds, J. W. (1980). The earthworm family Sparganophilidae (Annelida, Oligochaeta) in North America. *Megadrilogica*, 3(12), 189–204.

Reynolds, J. W. (1994). The distribution of earthworms (Annelida: Oligochaeta) in North America. *Megadrilogica*, 5(11), 73–129.

Reynolds, J. W. (2008a). The earthworms (Oligochaeta: Lumbricidae and Sparganophilidae) of Ontario, Canada. *Megadrilogica*, 12(12), 157–208.

Reynolds, J. W. (2008b). Sparganophilidae (Annelida, Oligochaeta) distribution in North America and elsewhere, revisited. *Megadrilogica*, 12(9), 125–143.

Reynolds, J. W. (2020). *Earthworms in American Ecoregions* (p. 433). Mauritius: LAP LAMBERT Academic Publishing.

Reynolds, J. W. (2022). The earthworms (Lumbricidae, Megascolecidae and Sparganophilidae) in Canada. *Canadian Food Inspection Agency*, Ottawa, i + 179 pp. ISBN: 978-0-660-44874-9.

Reynolds, J. W., & Wetzel, M. J. (2022). *Nomenclatura Oligochaetologica—a Catalogue of Names, Descriptions and Type Specimens* (Editio Secunda). Available from: https://nomenclatura-oligochaetologica.inhs.illinois.edu (Accessed 7 Novemebr 2022).

Righi, G. (1972). Contribuição ao conhecimento dos Oligochaeta brasileiros. *Papéis Avulsos de Zoologia*, 25, 148–166.

Righi, G. (1995). Colombian earthworms. In T. Van der Hammen, & A. G. Santos (Eds.), *Studies on Tropical Andean Ecosystems* (vol. 4, pp. 485–607). Berlin: Cramer.

Rodríguez, C., Borges, S., Martínez, M. A., Fragoso, C., James, S. W., & González, G. (2007). Biodiversidad y ecología de las lombrices de tierra em las islas caribeñas. In G. G. Brown, & C. Fragoso (Eds.), *Minhocas na América Latina: Biodiversidade e Ecologia* (pp. 79–98). Londrina: Embrapa.

Rodriguez-Seijo, A., Lourenço, J., Rocha-Santos, T. A. P., da Costa, J., Duarte, A. C., & Vala, H. (2017). Histopathological and molecular effects of microplastics in earthworms (*Eisenia andrei*) exposed to spiked soils. *Environmental Pollution*, 220, 495–503.

Römbke, J., Meller, M., & Garcia, M. (1999). Earthworm densities in central Amazonian primary and secondary forests and a polyculture forestry plantation. *Pedobiologia*, 43, 518–522.

Rosa, D. (1891). Die exotischen Terricolen des k. k. naturhistorischen Hofmuseums. Annalen des K. K. naturhistorischen Hofmuseums. *Wien*, 6, 379–406.

Rota, E., Benelli, S., Erséus, C., Soors, J., & Bartoli, M. (2018). New data and hypotheses on the invasiveness, habitat selection, and ecological role of the limicolous earthworm Sparganophilus tamesis Benham, 1892. *Fundamental and Applied Limnology*, 192(2), 129–136.

Rota, E., & Schmidt, O. (2006). Dichogaster bolaui (Oligochaeta: Octochaetidae), anunusual invader in a swimming pool in Ireland. *Journal of Natural History*, 40(3–4), 161–167.

Santos, A., Gorte, T., Demetrio, W. C., Ferreira, T., Nadolny, H., Cardoso, G. B. X., Tonetti, C., Ralisch, R., Nunes, A. P., Coqueiro, A. C. P., Leandro, H. C. L., Wandscheer, C. A. R., Bortoluzzi, J., Brown, G. G., & Bartz, M. L. C. (2018). Earthworm species in no-tillage agroecosystems and native Atlantic forests in Western Paraná, Brazil. *Zootaxa*, 4496, 517–534.

Shen, H., Barros, E., Mathieu, J., Hallaire, V., & Lavelle, P. (2011). Effects of the endogeic earthworm *Pontoscolex corethrurus* on soil infiltration capacity in a Brazilian clayey Ferralsol. *Applied Soil Ecology*, 47(2), 146–152.

Silva, E., Rosa, M. G., Schuhli, G. S., James, S. W., Decaëns, T., Bartz, M. L. C., Nadolny, H., Feijoo, A., & Brown, G. G. (2017). O potencial do DNA barcoding para a identificação e conservação de espécies de minhocas brasileiras. In F. M. S. Moreira, & M. C. M. Kasuya (Eds.), *Fertilidade e biologia do solo-Integração e tecnologia para todos* (vol. 2, pp. 549–570). Lavras: SBCS.

Sims, R. W. (1980). A classifications and the distribution of earthworms, suborder Lumbricina (Haplotaxida: Oligochaeta). *Bulletin of the British Museum Natural History (Zoology)*, 39(2), 103–124.

Sims, R. W. (1987). Review of the Central African earthworm family Eudrilidae (Oligochaeta). In A. M. Bonvicini Pagiliai, & P. Omodeo (Eds.), *On Earthworms* (pp. 359–388). Modena: Mucchi.

Sinha, R. K., Herat, S., Agarwal, S., Asadi, R., & Carretero, E. (2010). Vermiculture and waste management: Study of action of earthworms *Eisenia fetida*, *Eudrilus eugeniae* and *Perionyx excavatus* on biodegradation of some community wastes in India and Australia. *The Environmentalist*, 30(1), 34–44.

Six, J., Feller, C., Denef, K., Ogle, S. M., de Moraes Sa, J. C., & Albrecht, A. (2002). Soil organic matter, biota and aggregation in temperate and tropical soils – Effects of no-tillage. *Agronomie*, 22(7–8), 755–775.

Smiljkov, S., Trajanovski, S., & Budzakoska-Goreska, B. (2005). Biocenotic composition of the macrozoobenthos on different habitats from the littoral region of Lake Ohrid. *Prilozi*, 26(2), 143–155.

Sousa, S. C., Hernández-García, L. M., & Rousseau, G. X. (2020). A new species of Pontoscolex earthworm (Rhinodrilidae, Clitellata) from the Gurupi Biological Reserve, along with records of earthworm species from the Amazon region of Maranhão, Brazil. *Zootaxa*, 4801, 105–114.

Spurgeon, D. J., & Hopkin, S. P. (1996). Effects of metal-contaminated soils on the growth, sexual development, and reproduction of the earthworm *Eisenia fetida* with particular reference to zinc. *Ecotoxicology and Environmental Safety*, 35(1), 86–95.

Terhivuo, J. (1991). Dichogaster bolaui (Michaelsen) (Octochaetidae)—an allochthonous oligochaete invading urban sewer system. *Memoranda Societatis pro Fauna et Flora Fennica*, 67, 61–65.

Timm, T. (2020). Observations on the life cycles of aquatic Oligochaeta in aquaria. *Zoosymposia*, 17, 102–120.

Tondoh, J. E., & Lavelle, P. (2005). Population dynamics of Hyperiodrilus africanus (Oligochaeta, Eudrilidae) in Ivory Coast. *Journal of Tropical Ecology*, 21, 493–500.

Trakić, T., Valchovski, H., & Stojanović, M. (2016). Endemic earthworms (Oligochaeta: Lumbricidae) of the Balkan Peninsula: A review. *Zootaxa*, 4189, 251–274.

Tripathi, G., & Bhardwaj, P. (2004). Comparative studies on biomass production, life cycles, and composting efficiency of *Eisenia fetida* (Savigny) and *Lampito mauritii* (Kinberg). *Bioresource Technology*, 92(3), 275–283.

Valchovski, H. (2013). Contribution to the knowledge for distribution of Criodrilus lacuum (Annelida: Oligochaeta: Criodrilidae) from Bulgaria. *ZooNotes*, 44, 1–3.

van Groenigen, J. W., Lubbers, I. M., Vos, H. M. J., Brown, G. G., De Deyn, G. B., & van Groenigen, K. J. (2014). Earthworms increase plant production: A meta-analysis. *Scientific Reports*, 4, 6365. https://doi.org/10.1038/srep06365

Vejdovský, F. (1884). Beiträge zur Kenntniss der Oligochätenfauna Mitteleuropas. *Archiv für Naturgeschichte*, 50(1), 1–52.

Zicsi, A. (2007). Earthworm species, a searchable database. *Opuscula Zoologica*, 43(1), 97–99. https://doi.org/10.5281/zenodo.1045525

Zicsi, A., & Csuzdi, C. (2007). An annotated checklist of the earthworms of Chile (Oligochaeta). Earthworms from South America 43. In G. G. Brown, & C. Fragoso (Eds.), *Minhocas na América Latina: Biodiversidade e Ecologia* (pp. 235–245). Londrina: Embrapa.

Zicsi, A., Römbke, J., & Garcia, M. (2001). Regenwürmer (Oligochaeta) aus der Umgebung von Manaus (Amazonien). Regenwürmer aus Südamerika 32. *Revue Suisse de Zoologie*, 108(1), 153–164.

2 Terrestrial Earthworms of India

2.1 INTRODUCTION

Earthworms are well known as ecosystem engineers, their burrowing and feeding behaviors improving soil quality and creating favorable conditions for other organisms (Lavelle et al., 2006). Earthworms serve as bioindicators due to their responsiveness to environmental changes, making them valuable in environmental assessments and contamination studies (Howmiller & Beeton, 1971a,b; Julka, 1988a,b,c,d; Weber, 2007; Martins et al., 2008; Ozdemir et al., 2011; Pelosi & Röembke, 2016a,b; Velki & Ecimović, 2017). Globally, over 6000 earthworm species have been identified, with 3000–3500 considered valid (Csuzdi, 2012a,b,c; Reynolds & Wetzel, 2024a,b), and around 150 species with peregrine distributions (Blakemore, 2006a,b, 2012a,b). Extensive studies on earthworm biodiversity include recent findings (Mısırlıoğlu et al., 2023; Sautter et al., 2006a,b; Razafindrakoto et al., 2010; Rakotofiringa, & Blanchart, 2010). India's earthworm fauna is particularly diverse, comprising 459 species across ten families (Balakrishnan et al., 2024). Notable contributions to Indian earthworm taxonomy were made by Bourne (1886a,b,c), Beddard (1902a,b), Michaelsen (1907a,b, 1910a,b), Stephenson (1914–1923), Aiyer (1929), Gates (1937–1972a,b), Jamieson (1977a,b), and Julka (1976–1988). Julka & Paliwal (1993, 1994), including Moniligastridae, Lumbricidae, Almidae, and Megascolecidae (Narayanan et al., 2020a,b,c, 2021a,b, 2022; Ahmed et al., 2022a,b), and Narayanan et al. (2016–2023) have also contributed significant insights. In addition, Paliwal & Julka (2005), Verma & Yadav (2010), Yadav (2011, 2016), and others have provided detailed checklists for various regions. In India, new taxa continue to emerge, primarily from the Northeast and Western Ghats, adding to listings by Julka et al. (2004) and further expanded by Lone et al. (2020, 2022) and Tiwari et al. (2021). Endemism in India is high: around 71% of genera and 89% of species are unique to the region, while about 57 exotic species have been introduced, especially in union territories. India's earthworm fauna (Table 2.1, Figure 2.1) is well documented compared to other Asian countries (Figures 2.2–2.4).

TABLE 2.1
Family-Wise Earthworm Species

Family	Number of Species
Moniligastridae	88
Lumbricidae	17
Almidae	8
Rhinodrilidae	1
Acanthodrilidae	25
Eudrilidae	1

(Continued)

DOI: 10.1201/9781003633600-2

TABLE 2.1 *(Continued)*
Family-Wise Earthworm Species

Family	Number of Species
Ocnerodrilidae	16
Octochaetidae	133
Benhamiidae	5
Megascolecidae	155

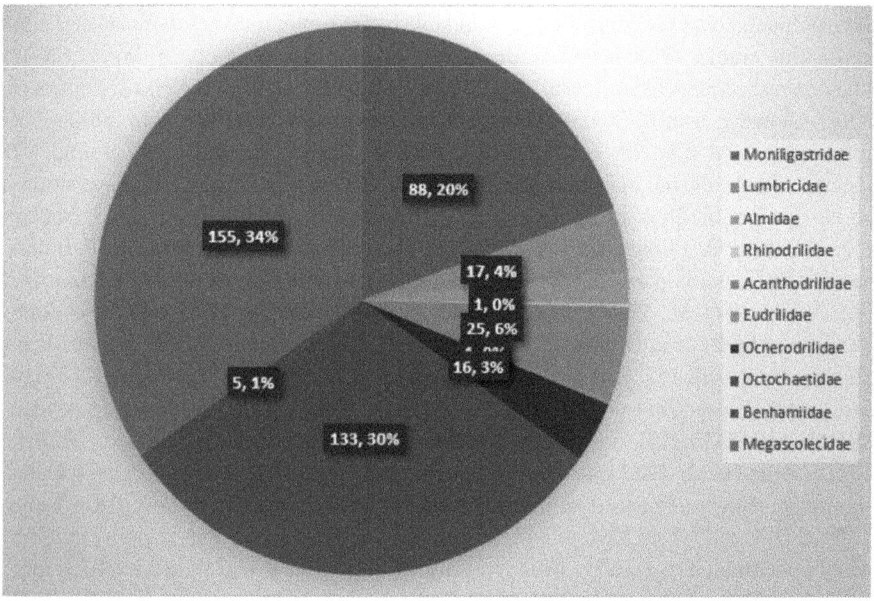

FIGURE 2.1 Family-wise distribution of earthworm species.

2.2 SYSTEMATIC ACCOUNT

Kingdom: Animalia
Phylum: Annelida
Class: Clitellata
Subclass: Oligochaeta
Order: Moniligastrida
1. Family: Moniligastridae (Claus, 1880)
Genera 3 and species 88
Order: Opisthopora
2. Acanthodrilidae Claus, 1880
Genera 2 and species 24
3. Almidae Duboscq, 1902
Genera 1 and species 7

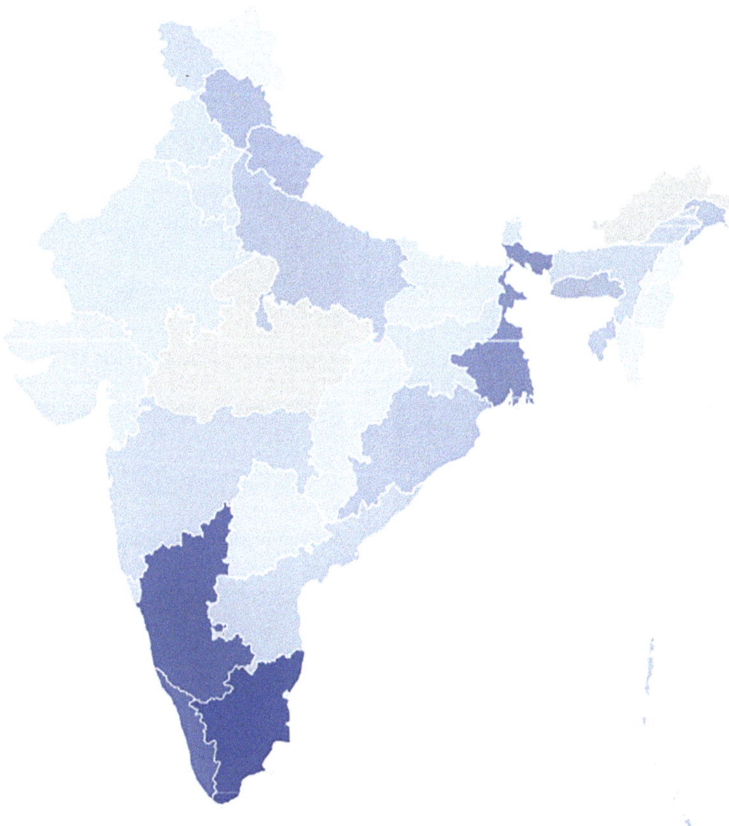

FIGURE 2.2 Map of India showing the species distribution from the viewpoint by country: the state-wise diversity of terrestrial earthworm species.

(Modified and updated from Mandal et al., 2023; Balakrishnan et al., 2024).

4. Benhamiidae Michaelsen, 1897
Genera 1 and species 5
5. Eudrilidae Claus, 1880
Genera 1 and species 1
6. Lumbricidae Rafinesque-Schmaltz, 1815
Genera 7 and species 16
7. Megascolecidae Rosa, 1891
Genera 14 and species 155
8. Ocnerodrilidae Beddard, 1891
Genera 15 and species 9
9. Rhinodrilidae Benham, 1879
Genera 1 and species 1
10. Octochaetidae Michaelsen, 1900.
Genera 22 and species 133

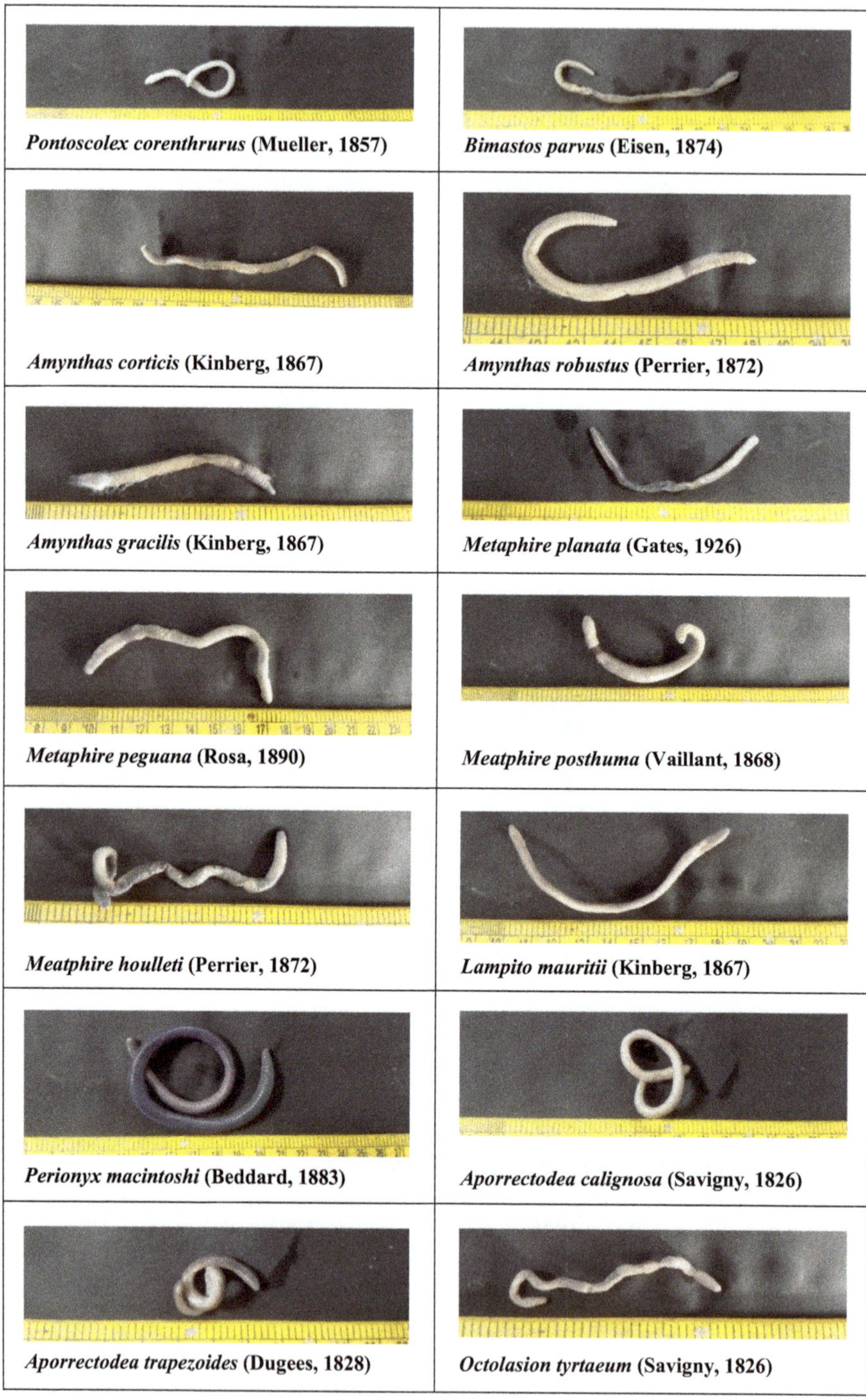

FIGURE 2.3 Earthworms from India.

(Image reproduced with permission of director, Zoological Survey of India).

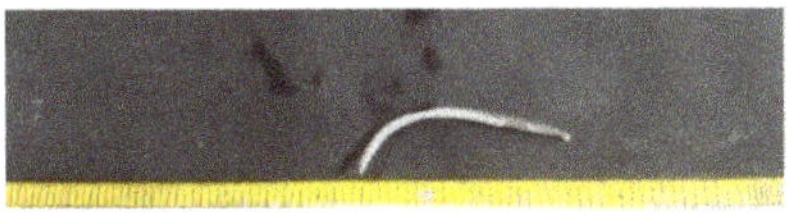

Metaphire californica (Kinberg, 1867)

Amynthas alexandri (Beddard, 1900)

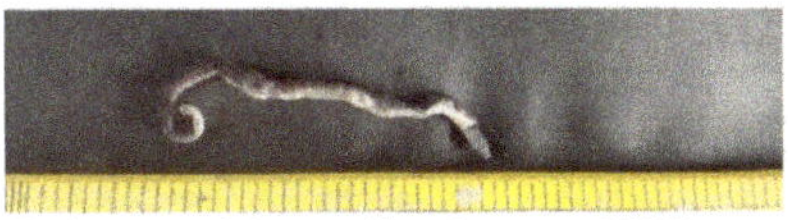

Perionyx pulvinatus (Stephenson, 1916)

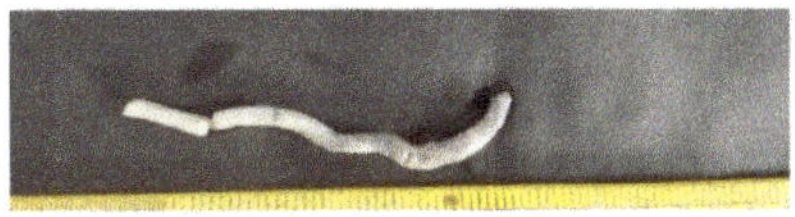

Drawida papillifer papillifer (Stephenson, 1917)

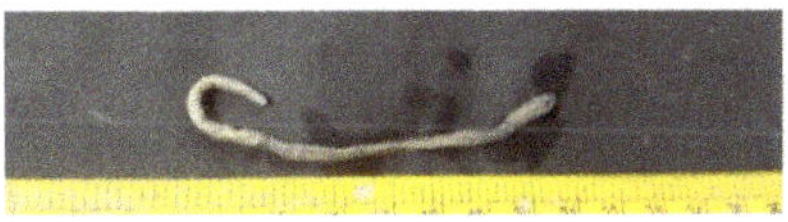

Drawida nepalensis (Michaelsen, 1907)

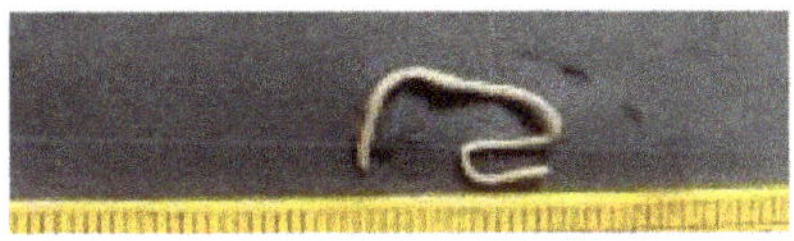

Eudrilus eugeniae (Kinberg, 1866)

Perionyx sansibaricus (Michaelsen,1891)

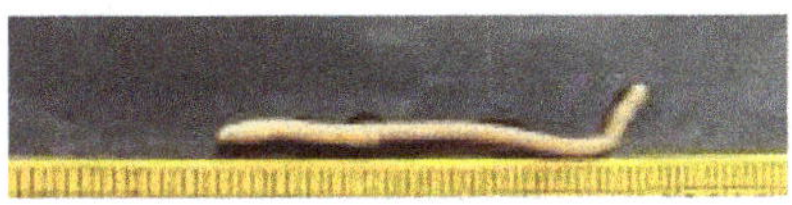

Eukerria kuekenthali (Michaelsen,1908)

Gordiodrilus elegans (Beddard, 1892)

Argilophilus macrochaetus macrochaetus (Stephenson, 1925)

Gordiodrilus elegans (Beddard, 1892)

Amynthas morrisi (Beddard, 1892)

Drawida limella (Gates, 1934)

Drawida willsi (Michaelsen, 1907)

FIGURE 2.3 (Continued)

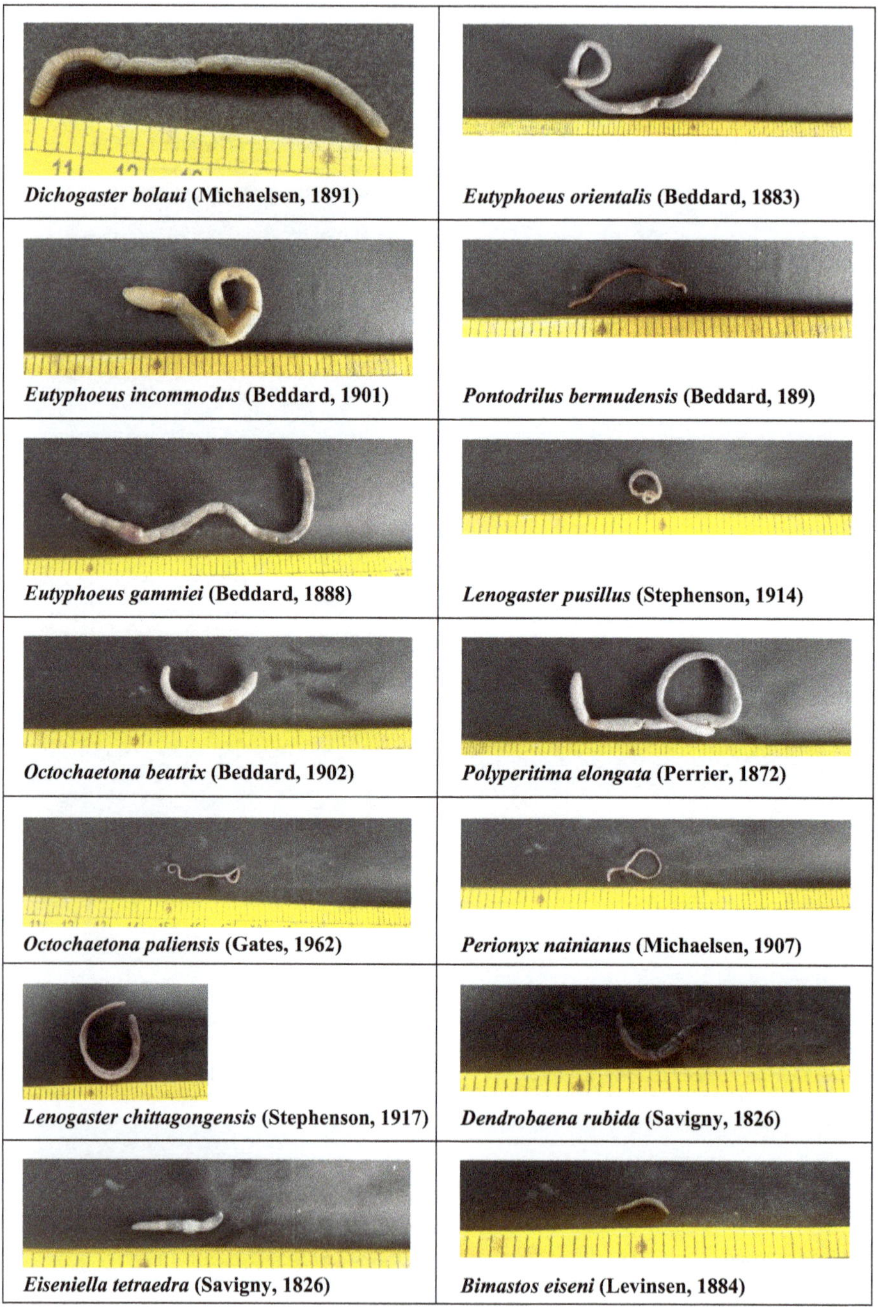

FIGURE 2.3 (Continued)

Metaphire anomola **(Michaelsen, 1907)**

Ocochaetona compta **(Gates, 1945)**

Pellogaster bengalensis **(Michaelsen, 1910)**

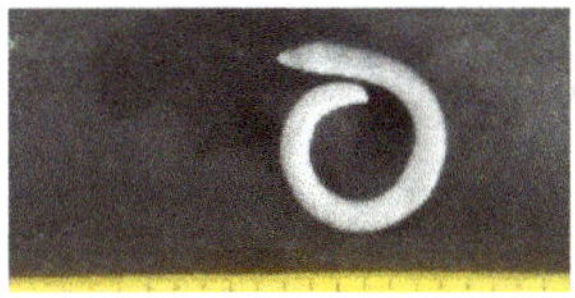

Drawida japanica **(Michaelsen, 1892)**

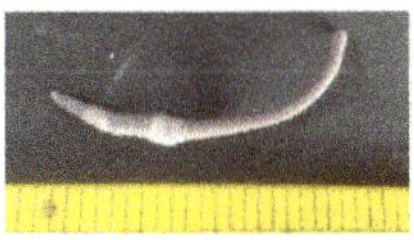

Eisenia fetida **(Savigny, 1826)**

Deccania alba **(Gates, 1949)**

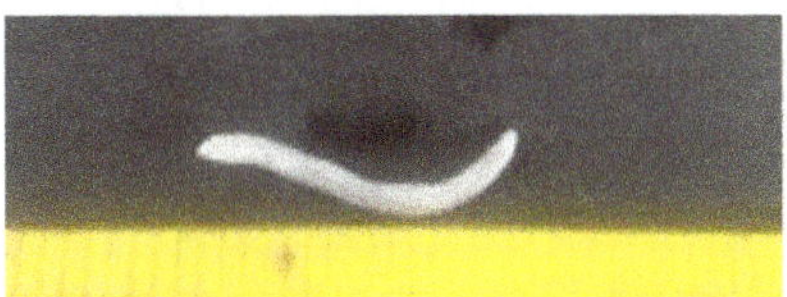

Drawida calebi **(Gates, 1945)**

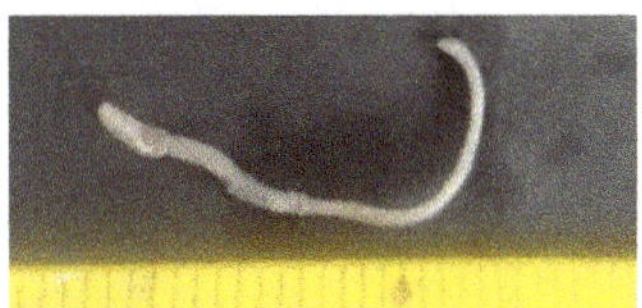

Dichogaster modigilianii **(Beddard, 1901)**

Ramiella sundargarhensis **(Julka, 1978)**

Eutyphoeus gammiei **(Beddard, 1888)**

Eutyphoeus nainianus **(Michaelsen, 1907)**

Glyphidrilus gangeticus **(Gates, 1958)**

Gordiodrilus ditheca **(Beddard, 1892)**

Ramiella bishambari **(Stephenson, 1914)**

FIGURE 2.3 (Continued)

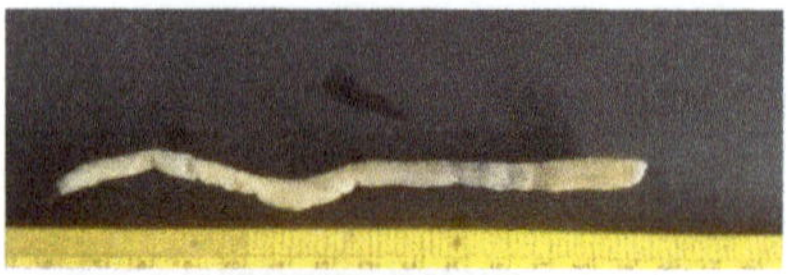

Eutyphoeus nicolsoni **(Beddard, 1901)**

Perionyx annandalei **(Michaelsen, 1907)**

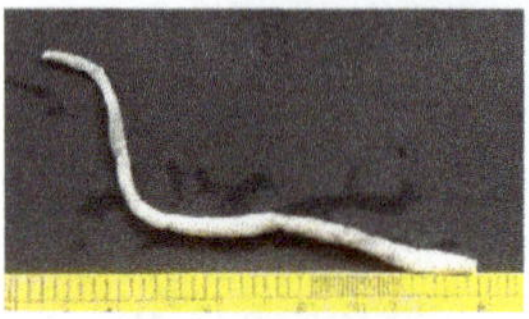

Tonscolex kalimpongensis
(Ahmed & Julka,2022)

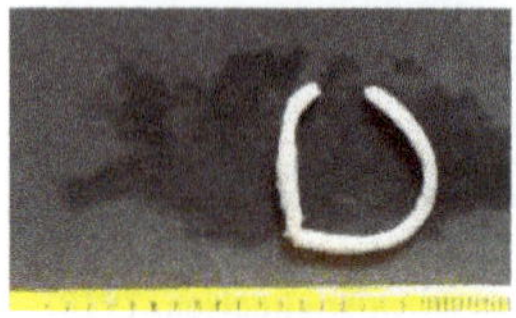

Dichogaster bolani **(Michaelsen, 1891)**

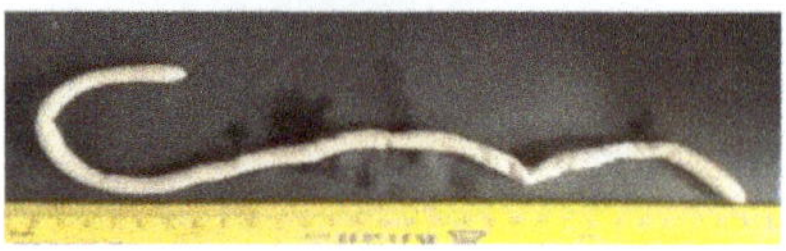

Megascolex konkanensis konkanensis
(Fedarb, 1898)

Dichogaster bolaui **(Michaelsen, 1991)**

Dichogaster affinis **(Michaelsen, 1890)**

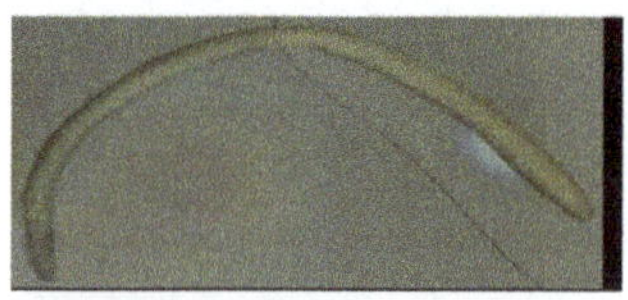

Perionyx fulvus **(Stephenson, 1916)**

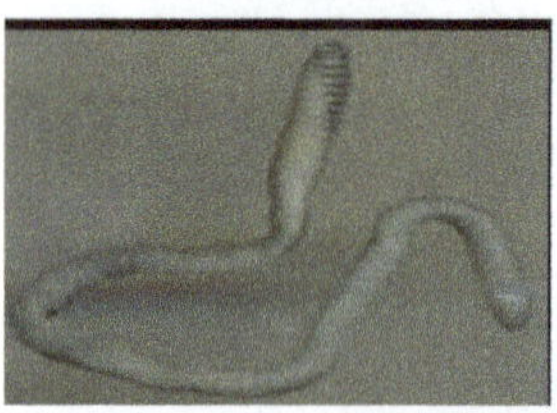

Perionyx jorpokriensis **(Julka, 1975)**

Eutyphoeus nepalensis **(Michaelsen, 1907)**

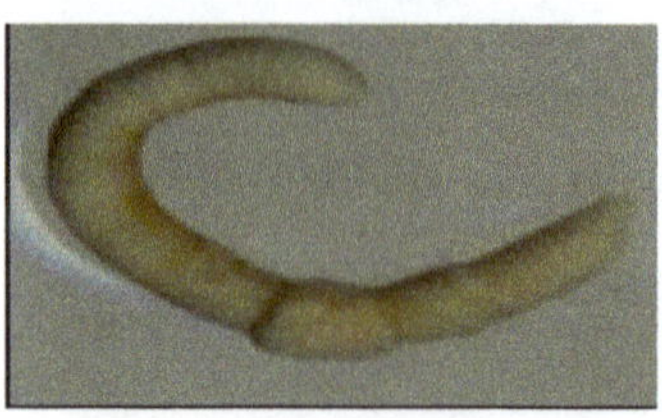

Eisenia hortensis **(Michaelsen, 1900)**

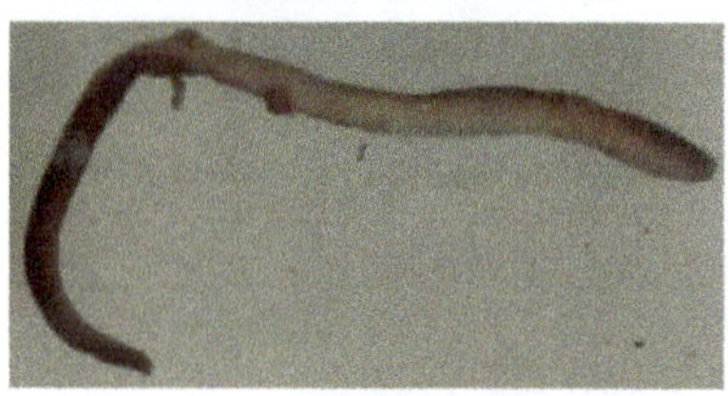

Perionyx variegatus **(Stephenson, 1923)**

FIGURE 2.3 (Continued)

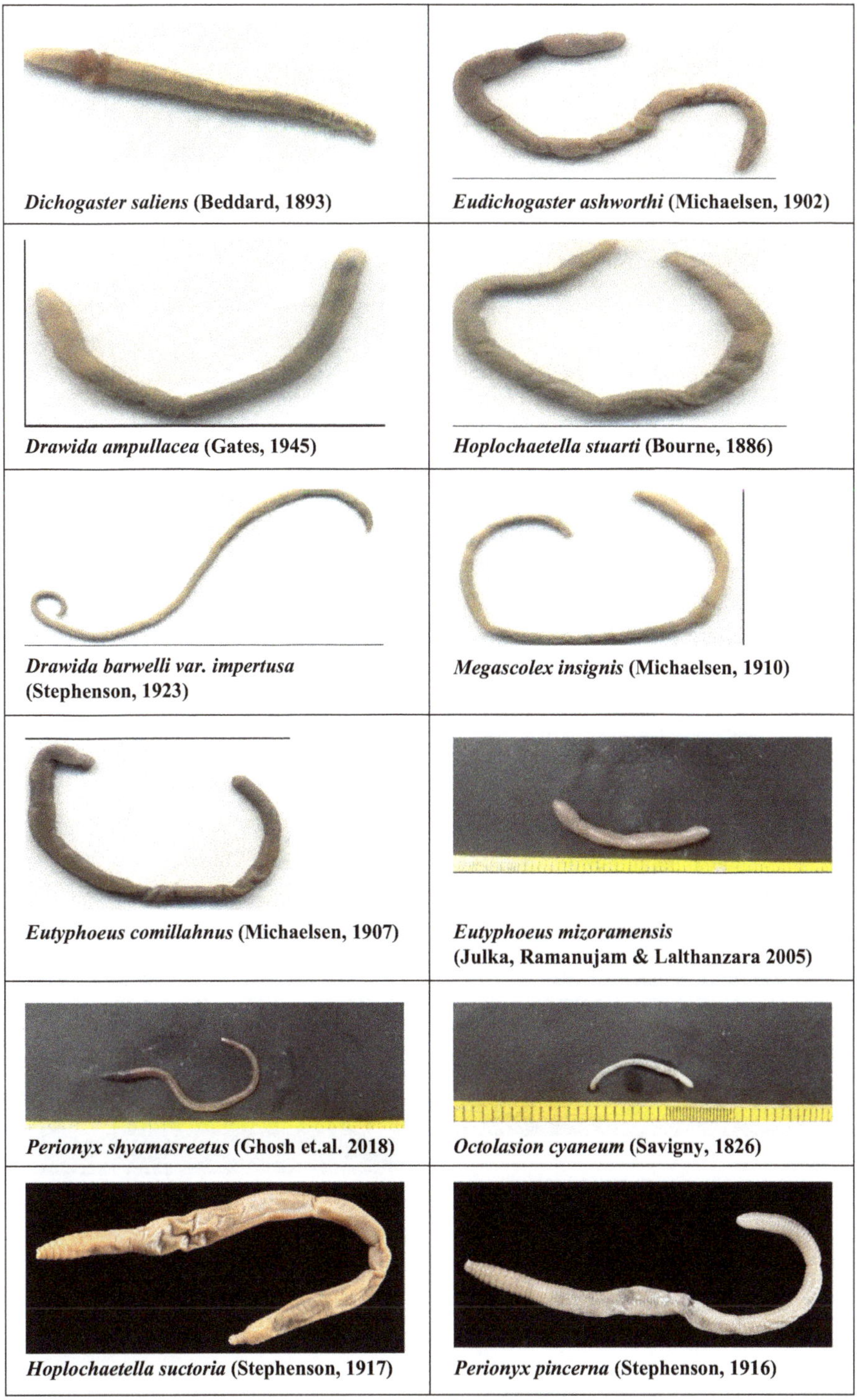

Dichogaster saliens (Beddard, 1893)

Eudichogaster ashworthi (Michaelsen, 1902)

Drawida ampullacea (Gates, 1945)

Hoplochaetella stuarti (Bourne, 1886)

Drawida barwelli var. impertusa (Stephenson, 1923)

Megascolex insignis (Michaelsen, 1910)

Eutyphoeus comillahnus (Michaelsen, 1907)

Eutyphoeus mizoramensis (Julka, Ramanujam & Lalthanzara 2005)

Perionyx shyamasreetus (Ghosh et.al. 2018)

Octolasion cyaneum (Savigny, 1826)

Hoplochaetella suctoria (Stephenson, 1917)

Perionyx pincerna (Stephenson, 1916)

FIGURE 2.3 (Continued)

Eutyphoeus foveatus (Rangoon, 18906)

Drawida bifida (Gates, 1945)

Drawida modesta (Stephenson, 1923)

Edugaster matherensis (R. J. Whittle, 2009)

Drawida longatria (Gates, 1925)

Eutyphoeus planatus (Gates, 1929)

Drawida scandens (Rao, 1921)

Drawida peguana (Gates, 1925)

Drawida constricta (Gates, 1929)

Glyphidrilus papillatus (Rosa, 1890)

Erythraeodrilus mullani (Stephenson, 1924)

Drawida burchardi (Michaelsen, 1903)

FIGURE 2.3 (Continued)

Eutyphoeus nicholsoni (Beddard, 1901)

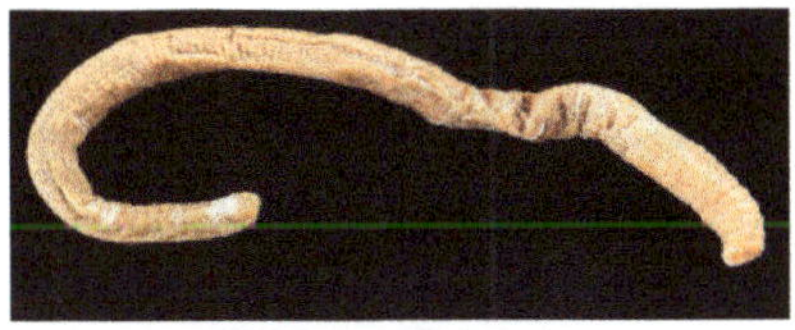

Erythraeodrilus khandalaensis (Stephenson, 1924)

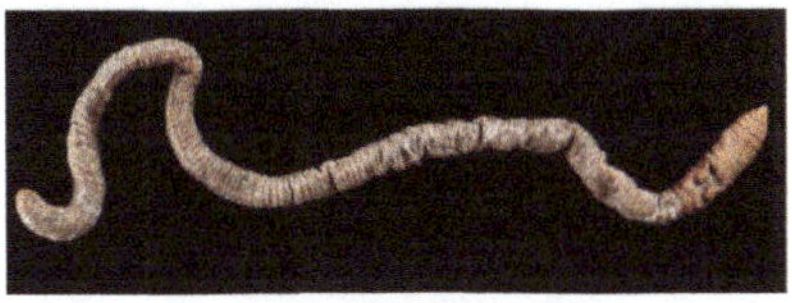

Drawida longatria (Gates, 1925)

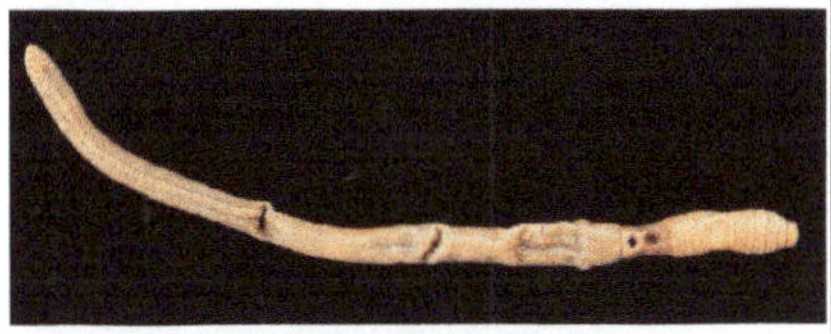

Eutyphoeus foveatus (Rosa, 1890)

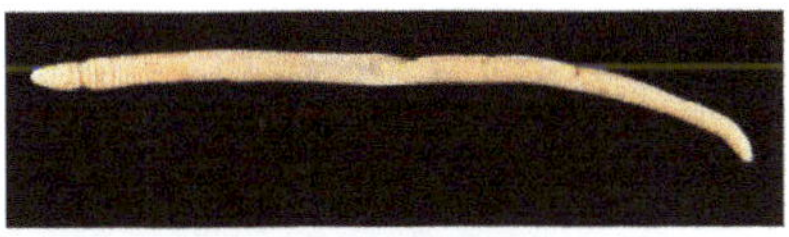

Drawida bifida (Gates, 1945)

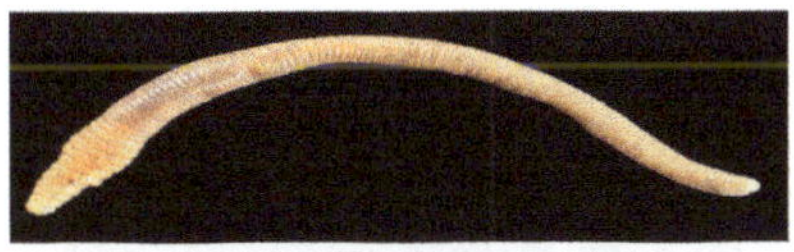

Drawida modesta (Stephenson, 1923)

Allolobophora jassyensis (Michaelsen, 1891)

Pheretima lignicola (Stephenson, 1914)

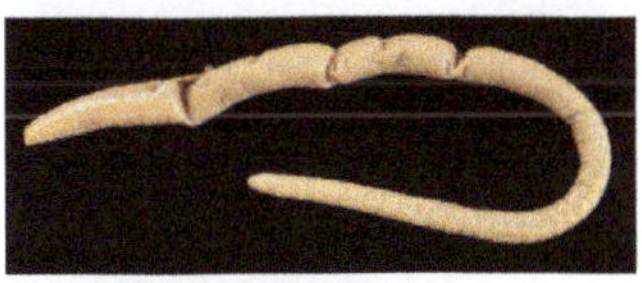

Hoplochaetella khandalaensis (Stephenson, 1924)

Hoplochaetella yearcaudensis (Rao, 1959)

Ramiella bishambari (Stephenson, 1914)

Lennogaster pusillus (Stephenson, 1920)

FIGURE 2.3 (Continued)

Desmogaster ferina **(Gates, 1943)**

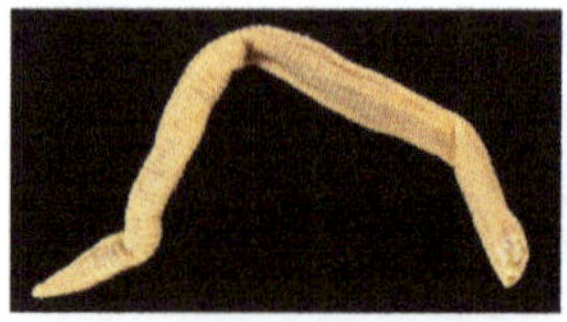

Perionyx millardi **(Stephenson, 1915)**

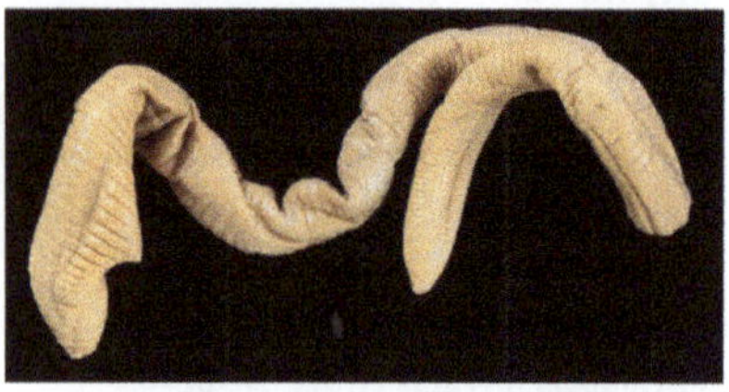

Drawida ghatensis **(Michaelsen, 1910)**

Drawida sulcata **(Michaelsen, 1907)**

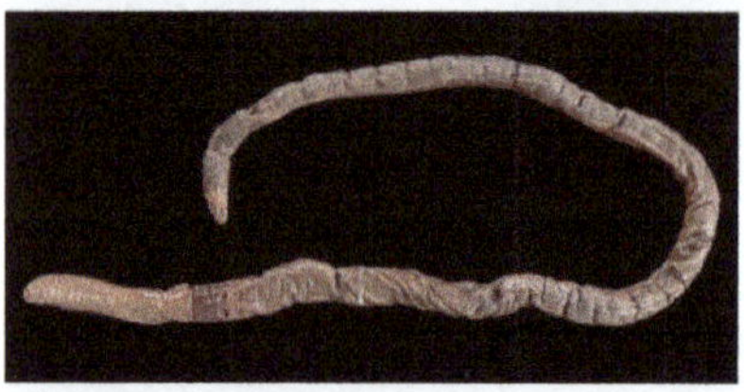

Eutyphoeus gigas **(Stephenson, 1917)**

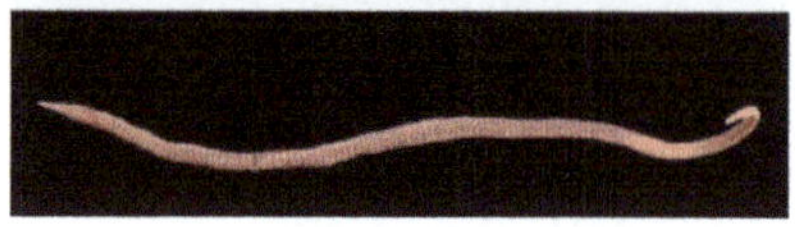

Perionyx gravelyi **(Stephenson, 1917)**

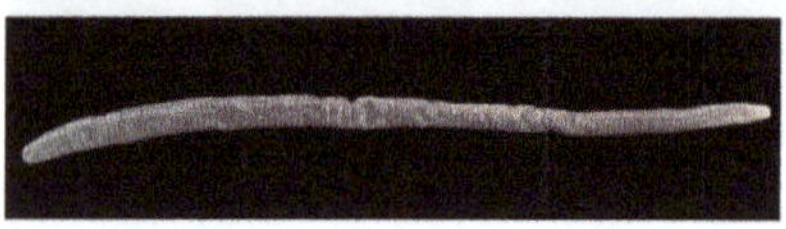

Perionyx pocrotensis
(Julka & Polmodal, 1925)

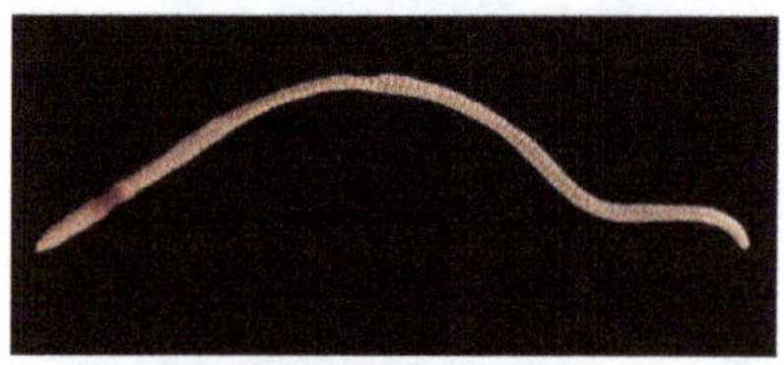

Perionyx simlensis **(Paliwal, 194)**

Tonoscolex michaelseni **(Julka, 1976)**

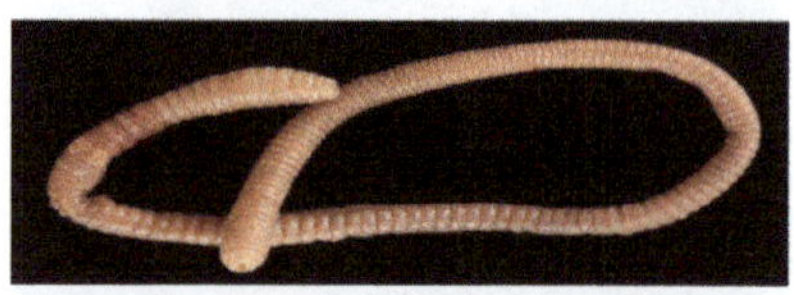

Notoscolex scutarius **(Michaelsen, 1907)**

Octochaetona parva **(Gates, 1962)**

Gordiodrilus elegans **(Beddard, 1892)**

FIGURE 2.3 (Continued)

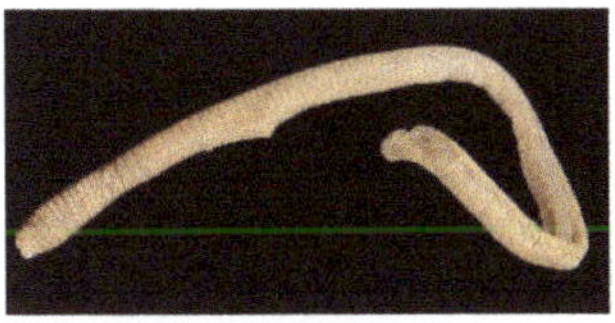

Drawida kempi **(Stephenson, 1914)**

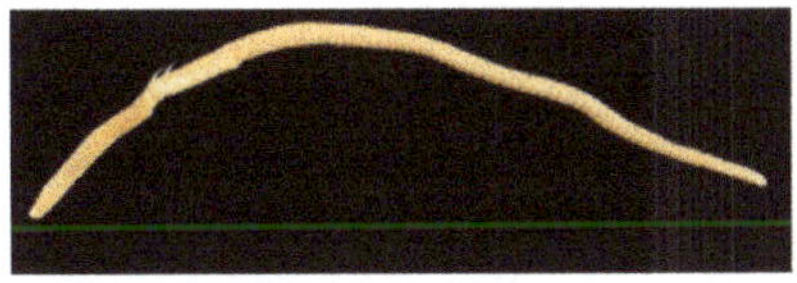

Malabaria levis **(Chen, 1938)**

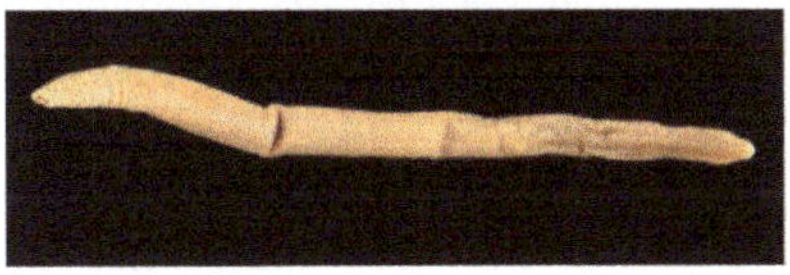

Drawida intermedia **(rao, 1921)**

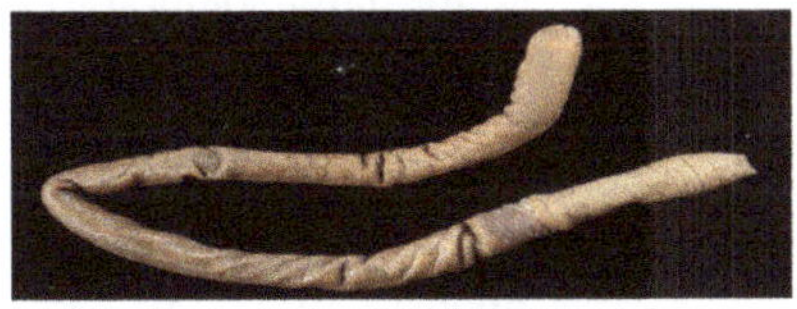

Eutyphoeus peguanus **(Gates, 1925)**

Tonoscolex depressus **(Gates, 1929)**

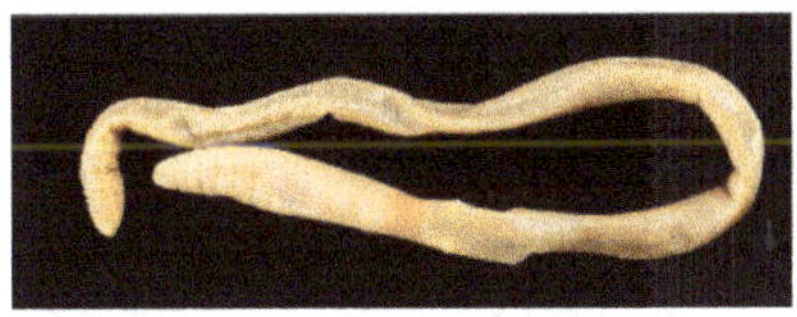

Pheretima heterochaeta **(Michaelsen, 1891)**

FIGURE 2.3 (Continued)

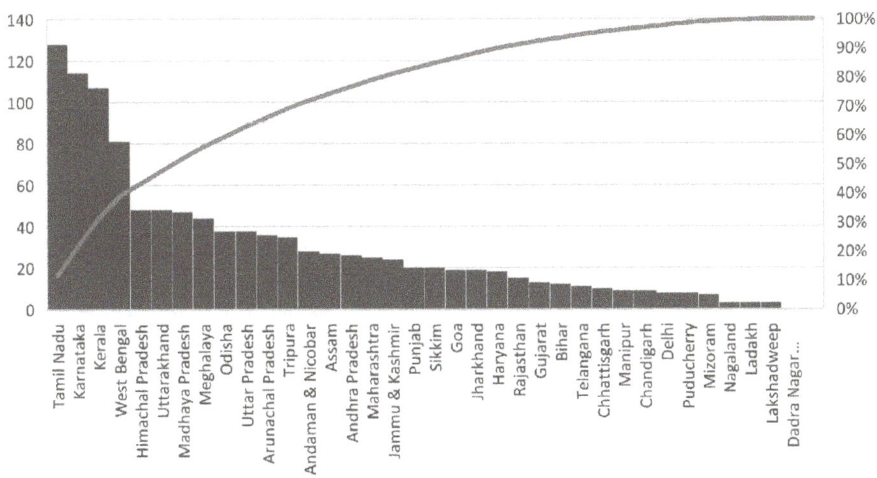

FIGURE 2.4 State-wise species distribution of terrestrial earthworms in India.
(Modified and updated from Mandal et al., 2023; Balakrishnan et al., 2024).

2.3 METHOD OF EARTHWORM SAMPLING

The techniques for sampling terrestrial earthworms in India vary widely, focusing on methods suited to different habitats and study objectives. A commonly used approach is hand sorting, which involves manual searching and collection of earthworms from soil samples, often enhanced by sorting through litter layers (Julka, 1988a, 1988b, 1988c, 1988d; Blakemore, 2000). Another method, formalin extraction, involves applying a dilute formalin solution to the soil to irritate earthworms, causing them to surface for easier collection; this technique has been widely used due to its effectiveness in capturing deep-burrowing species (Edwards & Lofty, 1977; Satchell, 1983). Quadrat sampling, where fixed area plots are sampled in detail, is also frequently employed to estimate earthworm density and diversity (Senapati et al., 1992; Bhattacharjee, 2002). Additionally, electro-shocking is a less common method where electrical currents are applied to the soil to stimulate earthworm emergence, although it is more costly and used selectively (Lee, 1985; Julka & Paliwal, 2005). In India, these techniques are often modified to account for soil type and moisture content.

2.4 IDENTIFICATION OF EXTERNAL CHARACTERISTICS OF EARTHWORMS

For earthworm morphology, researchers such as Julka (1988a, 1988b, 1988c, 1988d) and Stephenson (1923) emphasize examining form, shape, length, diameter, and the number of body segments, all essential for species identification. The color, which often shifts post-preservation, serves as a key characteristic (Gates, 1972a, 1972b, 1972c). Observing the prostomium type (prolobic, epilobic, zygolobic, or tanylobic) along with annuli counts in pre- and post-clitellar regions is important for classification (Julka & Paliwal, 2005). Detailed setal arrangement documentation, either perichaetine or lumbricine, and setal formula analysis further support taxonomic distinctions (Gates, 1972a, 1972b, 1972c). Align key reproductive pores spermathecal, male, female, and prostatic pores with setal positioning for accuracy (Julka, 1988a, 1988b, 1988c, 1988d). For observing pineal setae, critical to reproductive activity, methods involve removing and mounting them on slides in glycerine for detailed microscopic examination (Stephenson, 1923). Noting the presence or absence of genital markings like papillae, pits, and tubercula pubertatis also supports species identification (Julka & Paliwal, 2005). Additionally, recording clitellum structure, dorsal pores, nephridiopores, peristomium, and spermathecal pore features provides a comprehensive morphological profile, aiding in precise classification across Indian earthworm species.

2.5 IDENTIFICATION OF INTERNAL CHARACTERISTICS OF EARTHWORMS

In earthworms, the alimentary canal is a complex structure optimized for digestion and nutrient absorption, with key anatomical features well documented by studies such as Gates (1972a, 1972b, 1972c) and Julka (1988a, 1988b, 1988c, 1988d). The pharynx at the anterior end is followed by one or more muscular gizzards that mechanically break down ingested material. Calciferous glands, essential for calcium regulation, are situated along the canal and vary in number based on species (Stephenson, 1930). The intestine, starting at a specific segment, houses caecae and, where present, typhlosole structures that enhance surface area for efficient digestion and nutrient absorption (Julka & Paliwal, 2005). Circulation is supported by lateral hearts, which vary in number and extend to a terminal segment. The excretory system comprises nephridia (either holonephridia or meronephridia) that function along the body, according to specific species' adaptations (Kale, 1998). Earthworms' reproductive anatomy is notable, with paired testes, funnels, seminal vesicles, spermathecae, ovaries, oviducal funnels, and sometimes ovisacs positioned along the body, along with prostate glands that are either tubular or racemose. Together, the functions of the alimentary canal, reproductive, and excretory systems underscore earthworms' adaptations for soil habitats, promoting their role in soil health and ecosystem balance (Julka, 1988a, 1988b, 1988c, 1988d; Stephenson, 1930).

2.6 EARTHWORM PRESERVATION

The preservation methods for earthworms continue to follow techniques refined by various authors who have contributed extensively to specimen preservation protocols over the years. Authors such as Edwards and Bohlen (1996a, 1996b) and Martin and Lavelle (2000) emphasized that proper preservation begins with rinsing earthworms in water to remove soil and debris. According to these protocols, relaxation is achieved by immersing specimens in a mild solution, typically 70% ethanol, or a weak formalin solution (5–10%), as described by Lee (1985), to prevent contraction and maintain the natural form of the earthworms. This is followed by transferring the earthworms to a higher-concentration preservative, generally 70% ethanol, for long-term storage, as recommended by Rombke and Aira (2013). These established methods also suggest staining earthworms for anatomical studies and then re-preserving them in ethanol to enhance internal structure visibility. Proper labeling, including details such as collection date, location, and habitat, is highlighted by Jamieson (2000) as essential for future scientific accuracy and reference. These collective works form the foundation for contemporary preservation practices, enabling researchers to retain earthworm specimens in a state that supports detailed taxonomic and anatomical studies long after initial collection.

2.7 ENDEMIC TERRESTRIAL EARTHWORMS

The Indian subcontinent is one of the world's most biodiverse regions, characterized by a variety of ecological zones that range from the Himalayan mountains to the tropical rainforests of the Western Ghats and the fertile Indo-Gangetic plains. This diverse geography, combined with distinct climatic conditions, has contributed to the evolution of a unique and rich assemblage of terrestrial earthworm species, many of which are endemic to India (Julka, 1988a, 1988b, 1988c, 1988d; Julka & Paliwal, 1993a, 1993b). Earthworms, often referred to as "ecosystem engineers", play a crucial role in maintaining soil health, organic matter decomposition, and nutrient cycling, thereby influencing the productivity and stability of various ecosystems (Edwards & Bohlen, 1996a, 1996b). India's endemic earthworm fauna has been extensively studied by taxonomists and ecologists over the last century, with pioneering contributions from Stephenson (1923), Gates (1972a, 1972b, 1972c), and Julka (1988a, 1988b, 1988c, 1988d). Their studies have provided comprehensive taxonomic descriptions, ecological assessments, and distributional data on several endemic species. Endemic earthworms are often found in undisturbed habitats, such as primary forests, grasslands, and wetlands, where they play a vital role in maintaining ecological balance (Blakemore, 2008).

2.8 HISTORICAL PERSPECTIVES ON INDIAN EARTHWORMS

The earliest studies on Indian earthworms can be traced back to Beddard (1895) and Michaelsen (1900), who described several new species from the region. Later, Stephenson (1923) made significant contributions by publishing "The Fauna of British India: Oligochaeta", which remains a foundational work in Indian earthworm taxonomy. Gates (1972a, 1972b, 1972c) further refined the classification of Indian earthworms, recognizing many species as endemic to specific biogeographical zones. Julka (1988a, 1988b, 1988c, 1988d) conducted extensive surveys in India, particularly in the Western Ghats, and identified numerous new species belonging to the genera *Eutyphoeus, Drawida,* and *Perionyx.* The studies by Julka and Paliwal (1993, 1994) focused on endemic and rare species, emphasizing the importance of conservation measures. More recently, Narayanan et al. (2020a,b,c) and Senapati et al. (2022) have explored the ecological functions of endemic earthworms, particularly their roles in soil fertility and organic farming systems.

2.9 BIOGEOGRAPHICAL DISTRIBUTION OF INDIAN ENDEMIC EARTHWORMS

The endemic earthworm fauna of India is distributed across various biogeographical regions, each with distinct ecological conditions. The Western Ghats, a biodiversity hotspot with high rainfall and tropical forests, host unique genera such as *Eutyphoeus* and *Drawida* (Julka, 1988a, 1988b, 1988c, 1988d; Narayanan et al., 2020a,b,c). The Eastern Ghats, characterized by semi-evergreen forests and lateritic soils, support endemic species adapted to both dry and moist conditions (Gates, 1972a, 1972b, 1972c; Senapati et al., 2022). In the north-eastern Himalayas, high-altitude earthworm species exhibit cold tolerance and specialized burrowing adaptations (Stephenson, 1923; Julka & Paliwal, 1993a, 1993b). The Indo-Gangetic Plains, dominated by alluvial soils, are home to species like *Drawida nepalensis* (Stephenson, 1923; Blakemore, 2008). The Deccan Plateau, with its black and red soils, supports drought-resistant species, including certain *Eutyphoeus* species (Julka, 1988a,b,c,d).

2.10 ECOLOGICAL IMPORTANCE OF ENDEMIC EARTHWORMS

Endemic earthworms play a crucial role in maintaining soil health and ecosystem stability by actively participating in soil formation, organic matter decomposition, and nutrient cycling. Through their burrowing activities, they enhance soil aeration, improve water infiltration, and create microhabitats that support diverse microbial communities, thereby contributing to increased soil fertility and plant productivity

(Edwards & Bohlen, 1996a, 1996b; Senapati et al., 2022). Their feeding behavior accelerates the breakdown of organic matter, facilitating the release of essential nutrients such as nitrogen, phosphorus, and potassium, which are vital for plant growth (Julka, 1988a, 1988b, 1988c, 1988d; Blakemore, 2008). Some endemic species, such as *Perionyx excavatus* and *Eutyphoeus gammiei*, have been extensively studied for their potential applications in vermicomposting and sustainable agriculture, as they efficiently convert organic waste into nutrient-rich vermicast, improving soil structure and crop yield (Narayanan et al., 2020a,b,c). Furthermore, their role as bioindicators makes them valuable for assessing soil quality and environmental health, especially in regions facing habitat degradation and soil erosion (Gates, 1972a,b,c; Julka & Paliwal, 1993a, 1993b).

2.11 CONSERVATION AND THREATS TO ENDEMIC EARTHWORM SPECIES

Endemic earthworm species in India face significant threats due to habitat destruction, deforestation, agricultural expansion, urbanization, and climate change, which have collectively led to declining populations and habitat degradation (Julka, 1988a,b,c,d; Senapati et al., 2022). Habitat loss caused by deforestation and land conversion for agriculture disrupt the natural ecosystems that support these species, while soil pollution from pesticides and industrial activities reduces their reproductive success and habitat suitability (Gates, 1972a,b,c; Blakemore, 2008). Additionally, climate change–induced alterations in temperature and rainfall patterns further impact earthworm populations by disrupting soil moisture levels and food availability (Edwards & Bohlen, 1996a, 1996b). To mitigate these threats, conservation efforts must focus on establishing protected areas in biodiversity hotspots like the Western Ghats and north-eastern India, ensuring the preservation of endemic species and their habitats (Narayanan et al., 2020a,b,c). The promotion of sustainable agricultural practices, such as organic farming and reduced chemical pesticide use, can help maintain healthy earthworm populations while enhancing soil fertility (Julka & Paliwal, 1993a, 1993b). Furthermore, long-term ecological research and monitoring programs are crucial for understanding population dynamics, assessing threats, and formulating effective conservation strategies to protect these ecologically vital invertebrates (Senapati et al., 2022).

2.12 STUDY ON ENDEMIC EARTHWORMS IN INDIA

Investigate on endemic earthworms in India has significantly advanced our understanding of their taxonomy, ecology, and conservation status, with several notable contributions over the decades. Early taxonomic studies, such as those by Stephenson (1923), laid the foundation for identifying and classifying various earthworm species, including endemic genera like *Eutyphoeus* and *Drawida*. Later, Gates (1972a, 1972b, 1972c) and Julka (1988a, 1988b, 1988c, 1988d) provided further taxonomic revisions, refining species descriptions and distribution records. Ecological studies by Baskar et al. (2009) highlighted the crucial role of earthworms in soil formation, organic matter decomposition, and microbial interactions, emphasizing their contribution to soil fertility. More recent research has focused on their role in nutrient cycling and soil health, particularly in agroecosystems, where they enhance soil aeration, increase organic matter decomposition, and improve crop productivity (Edwards & Bohlen, 1996a, 1996b; Senapati et al., 2022). Studies on vermicomposting have explored the efficiency of species such as *Perionyx excavatus* and *Eutyphoeus gammiei* in converting organic waste into nutrient-rich compost, promoting sustainable agricultural practices (Narayanan et al., 2020a,b,c). Additionally, conservation-based research has assessed the impact of habitat destruction, pollution, and climate change on endemic earthworm populations, emphasizing the need for long-term ecological monitoring and biodiversity protection measures (Julka & Paliwal, 1993a, 1993b; Blakemore, 2008). The growing body of literature underscores the ecological significance of endemic earthworms and the urgent need to integrate them into soil conservation and sustainable land management strategies in India.

Discussion: India's endemic terrestrial earthworms are vital components of soil ecosystems, playing key roles in nutrient cycling, organic matter decomposition, and maintaining soil health. Their presence is crucial for sustaining agricultural productivity, biodiversity, and overall ecosystem stability. However, habitat destruction, deforestation, pollution, and climate change pose significant threats to their populations, necessitating urgent conservation efforts. Sustainable land management practices, including organic farming, habitat protection, and reduced chemical usage, can help preserve these ecologically essential invertebrates. Future research should focus on long-term monitoring, taxonomy, and ecological assessments to enhance conservation strategies and ensure the survival of endemic earthworm species. By integrating conservation efforts with sustainable agricultural policies, India can safeguard its rich earthworm diversity, promoting healthier soils and resilient ecosystems for future generations.

2.13 INVASIVE EARTHWORM SPECIES

Invasive earthworms, although beneficial in certain contexts, pose significant ecological challenges, particularly in regions where they were not historically present. The introduction of species like *Lumbricus terrestris* and *Amynthas agrestis* to North America has resulted in profound changes to soil dynamics, plant communities, and native invertebrate populations. For example, Bohlen et al. (2004) documented how the invasion of European earthworms into North American forests led to the depletion of organic matter and a decrease in native soil fauna. Similarly, Frelich et al. (2006) observed that earthworm invasion in temperate forests of the northern U.S. disrupted the structure of native plant communities by accelerating litter decomposition. The spread of these invasive species is often facilitated by human activities, including agriculture, horticulture, and bait fishing, which have been shown to significantly contribute to their distribution (Callaham et al., 2006). These studies underscore the importance of understanding and managing the spread of invasive earthworms to protect native ecosystems. Earthworms are the most familiar annelids to people (Rouse & Pleijel, 2007). Earthworms are characterized by hermaphroditism, reproductive organs in a limited number of segments, and direct development of eggs in a cocoon excreted by the clitellum—probable adaptations to life in a non-marine environment (Timm, 2012), even though parthenogenesis and regeneration are common in some species (Gates, 1972a,b,c). They occur naturally in all types of soils with sufficient food, moisture, and temperature, commonly occurring in grasslands, forests, agricultural fields, gardens, kitchen drainages, composting pits, and even in aquatic habitats and trees, except where recent volcanic activity, glaciations, inundation, or desertification preclude them (Blakemore, 2012a, 2012b, 2012c). Currently, 5738 species/subspecies of earthworms belonging to 23 families have been described (Mısırlıoğlu et al., 2023), with around 3000–3500 considered valid species (Csuzdi, 2012a,b,c). In India, approximately 463 species/subspecies from nine families have been reported (Narayanan et al., 2023a,b; Ahmed et al., 2023a, 2023b; Chang et al., 2024; Rodingpuia et al., 2024; Naik et al., 2024; Tiwari et al., 2024). The Western Ghats and western coastal plains are recognized as the regions with the highest earthworm species richness in India (Julka & Paliwal, 2005; Narayanan et al., 2023a,b). Earthworms move into new areas through active or passive dispersal methods (Edwards & Bohlen, 1996a, 1996b). Passive dispersal is common in peregrine or exotic species of earthworms, which mainly occur through anthropogenic activities and the transportation of cocoons in soil attached to the feet of birds or animals (Gates, 1972a, 1972b, 1972c; Julka, 1988a, 1988b, 1988c, 1988d, 2014; Blakemore, 2012a, 2012b, 2012c; Brown et al., 2006). The presence of various species of exotic or alien earthworms in different parts of the world was first detected in the early 20th century by Michaelsen (1903), who termed them "peregrine" species—species that have become established geographically over wide areas and outside their native ranges. These species are also called anthropochorous, allochthonous, or cosmopolitan species (Blakemore, 2012a, 2012b, 2012c). Currently, around 150 species are considered peregrine globally (Blakemore, 2012a, 2012b, 2012c). In India, the Western Ghats and western coastal plains account for approximately 58.4% of the country's earthworm diversity (Narayanan et al., 2020a,b,c). Taxonomic studies on the oligochaetes of the Western Ghats began at the end of the 19th century with the description of *Perichaeta lawsoni* (= *Megascolex lawsoni*) from Nilgiris (Tamil Nadu) by Bourne (1886a, 1886b). Recently, around 271 species are reported from the Western Ghats (Narayanan et al., 2020a,b,c, 2021a,b, 2022a,b, 2023a, 2023b, 2024a). This includes a significant number of alien species from different zoogeographical realms, including species introduced from other parts of Asia. Bourne (1886a,b,c) also described *Perichaeta mirabilis* from Naduvatam, which is now considered a synonym of *Amynthas corticis* (Kinberg, 1867), thus making it one of the first alien earthworm species recorded from the Western Ghats. However, the diversity and distribution pattern of alien earthworm species in the region are still not fully understood. This work aims to present a checklist of all the valid alien earthworm species recorded from the Western Ghats region.

Invasive earthworms primarily belong to the families Lumbricidae, Megascolecidae, and Acanthodrilidae and include several species that have been well documented for their ecological impact. For example, *Lumbricus terrestris*, a deep-burrowing (anecic) species from the Lumbricidae family, has been introduced to North America and Asia, where it disrupts soil structure and plant communities (Hendrix & Bohlen, 2002). Other invasive European earthworms such as *Aporrectodea* spp. are commonly found in disturbed soils and agricultural fields, while *Dendrobaena octaedra*, a litter-dwelling (epigeic) species, has high invasion potential due to its ability to rapidly decompose organic matter (Frelich et al., 2006). In the family Megascolecidae, *Amynthas agrestis*, also known as the "jumping worm", depletes organic matter at alarming rates, significantly altering ecosystems (Frelich et al., 2006). Additionally, species like *Pheretima* spp., known for their burrowing habits, further contribute to soil structure alteration. From South America and Africa, *Pontoscolex corethrurus* and *Eudrilus eugeniae*, the latter being a globally distributed composting worm, also exemplify invasive species, causing homogenization of soils and alterations in nutrient cycles (Hendrix & Bohlen, 2002). These species have been extensively studied for their significant ecological impact and their role in disrupting native ecosystems (Hendrix & Bohlen, 2002; Frelich et al., 2006).

2.14 IMPACT ON NATIVE BIODIVERSITY

Invasive earthworms have significant effects on various aspects of ecosystem function, particularly on soil fauna, plant communities, and microbial diversity. Studies by Bohlen et al. (2004) have shown that European earthworms introduced to North America disrupt native soil invertebrate communities by outcompeting local species such as millipedes, beetles, and fungi-dependent organisms, ultimately reducing biodiversity. In forests, invasive earthworms accelerate the decomposition of leaf litter, depleting nutrients and diminishing plant diversity, as observed by Hale et al. (2005). In grasslands and agricultural systems, species like *Amynthas agrestis* negatively impact crop growth by altering root patterns, further disrupting the ecosystem (Greiner et al., 2012). Additionally, the invasion of earthworms has been shown to affect microbial communities by shifting fungal and bacterial compositions, which results in lower nitrogen retention in soils (Burtelow et al., 1998; Ferlian et al., 2018). These disruptions demonstrate how invasive earthworms can profoundly alter ecosystem dynamics across multiple trophic levels.

2.15 ECOLOGICAL CONSEQUENCES OF EARTHWORM INVASIONS

Invasive earthworms significantly disrupt soil structure, nutrient cycling, and local biodiversity. For instance, species like *Pontoscolex corethrurus* alter soil porosity and hydrology, leading to nutrient leaching, as reported by Blouin et al. (2013). This disruption is also evident in forested areas, where the absence of earthworm-resistant vegetation contributes to increased erosion rates (Suárez et al., 2006). The changes in soil structure are accompanied by alterations in nutrient cycling, with enhanced decomposition rates causing a rapid loss of organic carbon (Eisenhauer et al., 2007). Furthermore, high bioturbation rates from invasive earthworms contribute to increased CO_2 emissions from soils (Zhang et al., 2020). In addition to these impacts, invasive species like *Amynthas* in tropical regions outcompete indigenous earthworms due to their faster reproduction rates, as highlighted by James & Hendrix (2004). The ecological consequences extend beyond soil dynamics; the loss of moist organic layers caused by earthworm invasions affects amphibians such as salamanders (Maerz et al., 2009), while a decline in invertebrate prey species leads to reduced populations of ground-nesting birds (Loss et al., 2013). These studies demonstrate how invasive earthworms can have cascading effects on both abiotic and biotic components of ecosystems.

2.16 MANAGEMENT STRATEGIES FOR INVASIVE EARTHWORMS

Preventing the spread of invasive earthworms and mitigating their ecological impact requires a combination of early detection, physical control, and soil management techniques. Monitoring bait use is crucial, as invasive worms are often spread through fishing bait disposal, a fact highlighted by Callaham et al. (2006). Regulations to restrict the sale of invasive species are necessary to curb this pathway. Additionally, soil quarantine measures are important, as nursery plants and potted soils can harbor invasive species (Gundale et al., 2005). Early detection is enhanced through DNA barcoding using COI mitochondrial markers, as demonstrated by Chang et al. (2021). In terms of control, physical methods such as hand removal of worms like *Amynthas agrestis* in forestry areas can help reduce their populations (Görres et al., 2014). Predation by natural enemies, including predatory beetles and nematodes, is also being explored as a biocontrol option (Zhang et al., 2019). Soil management techniques such as altering soil pH and moisture can make conditions less favorable for invasive worms, with acidification reducing their viability (Lubbers et al., 2017). Mulching and retaining leaf litter can help prevent the rapid loss of organic matter in invaded ecosystems (Fisichelli et al., 2012), while the use of allelopathic plants to create biological barriers offers a promising strategy to deter worms (Nouri-Aiin et al., 2020). Together, these strategies form a comprehensive approach to managing invasive earthworm populations.

2.17 DISCUSSION

The earthworm fauna of India span 28 states and eight Union Territories, totaling 459 species. This includes a variety of species distributed as follows: Andhra Pradesh (26 species), Arunachal Pradesh (36 species), Assam (27 species), Bihar (12 species), Chhattisgarh (10 species), Gujarat (13 species), Goa (19 species), Haryana (18 species), Himachal Pradesh (48 species), Jharkhand (19 species), Karnataka (114 species), Kerala (107 species), Madhya Pradesh (47 species), Maharashtra (25 species), Manipur (9 species), Meghalaya (44 species), Mizoram (7 species), Nagaland (3 species), Odisha (38 species), Punjab (20 species), Rajasthan (15 species), Sikkim (20 species), Tamil Nadu (128 species), Telangana (11 species), Tripura (35 species), Uttar Pradesh (38 species), Uttarakhand (48 species), West Bengal (81 species), Andaman & Nicobar (28 species), Chandigarh (9 species), Dadra Nagar Haveli, Daman & Diu (0 species), Delhi (8 species), Jammu & Kashmir (24 species), Ladakh (2 species), Lakshadweep (3 species), and Puducherry (3 species). This extensive distribution highlights the rich biodiversity of earthworms in India, contributing significantly to the understanding of soil health and ecological balance across the nation.

Invasive earthworms present significant challenges to ecosystems worldwide, particularly in regions where they are not native. Their introduction, often through human activities, has led to profound disruptions in soil dynamics, plant communities, and native biodiversity. The impacts are multifaceted, ranging from the alteration of soil structure and nutrient cycles to the decline of native species and the alteration of microbial communities. These changes can result in reduced soil fertility, increased erosion, and the loss of biodiversity, affecting both terrestrial and aquatic ecosystems. Furthermore, the spread of these species is often facilitated by activities such as agriculture, horticulture, and fishing, which underscores the need for better management and prevention strategies. Effective control measures must combine early detection, physical removal, and soil management techniques, along with the exploration of biological control options. Addressing the issue of invasive earthworms requires a comprehensive and integrated approach that not only aims to control their spread but also seeks to restore the affected ecosystems. This calls for continued research and collaboration to better understand the ecological consequences of earthworm invasions and to develop sustainable solutions that protect native biodiversity and ecosystem services.

REFERENCES

Ahmed, M., Ahmad, S., & Bhat, S. A. (2022). Recent updates on the biodiversity of Indian earthworms. *Soil Biology and Biochemistry*, 108, 119–130.

Ahmed, S., Julka, J. M., Banerjee, D., & Marimuthu, N. (2023a). A new speciesof the genus Hoplochaetella Michaelsen 1900 (Clitellata: Octochaetidae) from the Deccan Peninsula Biogeographic Zone, India. *Zootaxa*, 5346(2), 173–185.

Ahmed, S., Julka, J. M., Banerjee, D., & Marimuthu, N. (2023b). A new speciesof the genus Eutyphoeus Michaelsen 1900 (Clitellata: Acanthodrilidae) from the North-Eastern Biogeographic Zone, India. *Zootaxa*, 5380(2), 167–172.

Aiyer, P. S. (1929). The Indian earthworms. *The Journal of the Bombay Natural History Society*, 33, 31–42.

Balakrishnan, S., Mondal, T., Mandal, C. K., Ahmed, S., Kumar, Y., Sivaleela, J. S. G., & Raghunathan, C. (2024). *Fauna of India Checklist, Annelida: Polychaete: Sipuncula: Clitella: Hirudinida: Crassiclitellata: Moniligastrida.* Online Version 1.0. Zoological Survey of India. https://doi.org/10.26515/Fauna/1/2023/Annelida: Polychaeta, Sipuncula, Clitella, Hirudinida, Crassiclitellata, Moniligastrida

Baskar, M., Balamurugan, M., & Gunasekaran, S. (2009). Ecological studies on the role of earthworms in soil formation, organic matter decomposition and microbial interactions emphasizing their contribution to soil fertility. In: N. Munuswamy, S. Gunasekaran, A. S. Nair, & S. M. Krishnan (Eds.), *Proceedings of the 7th International Symposium on Earthworm Ecology* (pp. 48–52 or similar range, the snippet just mentions p. 48). Chennai, India: ISMRC publisher/printer.

Beddard, F. E. (1895). *A Monograph of the Order of Oligochaeta.* Oxford: Clarendon Press.

Beddard, F. E. (1902a). *Earthworms and Their Allies.* Cambridge University Press.

Beddard, F. E. (1902b). The earthworms of India. *Annals and Magazine of Natural History*, 8, 323–340.

Bhattacharjee, G., & Chaudhuri, P. S. (2002). Cocoon production, morphology, hatching pattern and fecundity in seven tropical earthworm species—a laboratory based investigation. *Journal of Biosciences*, 27, 283–294.

Blakemore, R. J. (2000). Ecology of earthworms under the 'Haughley Experiment' of organic and conventional management regimes. *Biological Agriculture & Horticulture*, 18(2), 141–159.

Blakemore, R. J. (2006a). *Cosmopolitan Earthworms—an Eco-Taxonomic Guide to the Peregrine Species of the World.* Kippax, Australia: VermEcology.

Blakemore, R. J. (2006b). Earthworm biodiversity and conservation in Australia. *Soil Biology and Biochemistry*, 38, 1832–1841.
Blakemore, R. J. (2008). *Cosmopolitan Earthworms—an Eco-Taxonomic Guide to the Peregrine Species of the World*. VermEcology.
Blakemore, R. J. (2012a). *Cosmopolitan Earthworms—an Eco-Taxonomic Guide to the Peregrine Species of the World* (5th ed.). Yokohama, Japan: VermEcology.
Blakemore, R. J. (2012b). Revised earthworm taxonomy of Australian species and genera. *Soil Organisms*, 84, 115–133.
Blakemore, R. J. (2012c). *Earthworms of the World, Volume 1: Ecology and Distribution* (pp. 1–385). Springer.
Blouin, M., Hodson, M. E., Delgado, E. A., & Brussaard, L. (2013). A review of earthworm impact on soil function and ecosystem services. *European Journal of Soil Science*, 64(2), 161–182. https://doi.org/10.1111/ejss.12025
Bohlen, P. J., Parmelee, R. W., & Blair, J. M. (2004). Integrating the effects of earthworms on nutrient cycling across spatial and temporal scales. In C. A. Edwards (Ed.), *Earthworm Ecology* (pp. 161–180). Boca Raton, FL: CRC Press. http://dx.doi.org/10.1201/9781420039719.ch9
Bourne, A. G. (1886a). An account of the earthworms collected by the Percy Sladen Expedition to the Seychelles in 1905. *Proceedings of the Zoological Society of London*, 1886, 331–342.
Bourne, A. G. (1886b). Description of Perichaeta lawsoni from Nilgiris. *Journal of the Bombay Natural History Society*, 1(4), 224–226.
Bourne, G. H. (1886c). The earthworms of India and their economic significance. *Proceedings of the Zoological Society of London*, 1886, 252–265.
Brown, G. G., Pashanasi, B., Villenave, C., Patron, J. C., Senapati, B. K., Giri, S., Martins, F. A., & Araujo, A. C. (2006). The role of earthworms in the ecology of tropical and subtropical ecosystems. *Soil Biology and Biochemistry,* 38(11), 2923–2932. https://doi.org/10.1016/j.soilbio.2006.05.013
Burtelow, A. E., Bohlen, P. J., & Groffman, P. M. (1998). Soil fauna and microbial dynamics in agricultural soils. *Soil Science Society of America Journal*, 62(4), 1022–1031. https://doi.org/10.2136/sssaj1998.03615995006200040024x
Callaham, M. A., Jr., James, S. W., Snyder, B. A., & Oberg, E. T. (2006). Human-mediated dispersal of invasive earthworms. In P. F. Hendrix (Ed.), *Earthworm Ecology* (2nd ed., pp. 232–248). CRC Press.
Chang, C.-H., Snyder, B. A., & Szlavecz, K. (2021). DNA barcoding of invasive earthworms in tropical regions. *Molecular Ecology*, 30(8), 2215–2224. https://doi.org/10.1111/mec.15873
Chang, D. F., Li, Y. B., & Zhang, J. H. (2024). New findings on earthworm species in tropical soils. *Journal of Soil Biology*, 45(2), 109–115.
Claus, C. (1880). Grundzüge der Zoologie. Zum Gebrauche an Hochschulen und höheren Lehranstalten. *Vierte verbesserte und vermehrte Auflage*. Erster Band. Marburg: N. G. Elwert'sche Verlagsbuchhandlung. viii + 822 pp.
Csuzdi, C. (2012a). Earthworm biodiversity in the world. *Soil Organisms*, 84, 3–13.
Csuzdi, C. (2012b). Earthworm species of the world: A review of known species and families. In *Earthworm Ecology and Soil Biology* (pp. 1–19). Springer.
Csuzdi, C. (2012c). Earthworm species, a searchable database. *Opuscula Zoologica Budapest*, 43(1), 97–99.
Edwards, C. A., & Lofty, J. R. (1977). *Biology of Earthworms (2nd ed.)*. Chapman & Hall, London, UK.
Edwards, C. A., & Bohlen, P. J. (1996a). *Biology and Ecology of Earthworms*. London: Chapman & Hall.

Edwards, C. A., & Bohlen, P. J. (1996b). *Biology and Ecology of Earthworms* (pp. 1–331). Springer.

Eisenhauer, N., Partsch, S., Parkinson, D., & Scheu, S. (2007). Earthworm invasion in North America: Food web effects and impacts on soil carbon. *Biological Invasions*, 9(1), 75–89. https://doi.org/10.1007/s10530-006-9009-5

Ferlian, O., Eisenhauer, N., Aguirrebengoa, M., Camara, M., Ramirez-Rojas, I., Santos, F., Tanalgo, K., & Thakur, M. P. (2018). Invasive earthworms on soil microorganisms: Temperature and moisture modulate the relative importance of direct and indirect effects. *Soil Biology and Biochemistry*, 116, 105–113. https://doi.org/10.1016/j.soilbio.2017.10.005

Fisichelli, N. A., Frelich, L. E., & Reich, P. B. (2012). Sapling growth responses to warmer temperatures 'cooled' by browse pressure. *Global Change Biology*, 18(11), 3455–3463. https://doi.org/10.1111/j.1365-2486.2012.02785.x

Frelich, L. E., Hale, C. M., Scheu, S., Holdsworth, A. R., Heneghan, L., Bohlen, P. J., & Reich, P. B. (2006). Earthworm invasion into previously earthworm-free temperate and boreal forests. *Biological Invasions*, 8(6), 1235–1245. https://doi.org/10.1007/s10530-006-9019-3

Gates, G. E. (1937–1972a). Earthworms of the world. *Transactions of Connecticut Academy of Arts and Sciences*, 39, 1–246.

Gates, G. E. (1937–1972b). On some earthworms from India. *Records of the Indian Museum*, 39, 213–257.

Gates, G. E. (1972a). Burmese earthworms, an introduction to the systematics and biology of Megadrile Oligochaetes with special reference to Southeast Asia. *Transactions of the American Philosophical Society*, 62(7), 1–326.

Gates, G. E. (1972b). Earthworms of the genus *Acanthodrilus* in the neotropics. *Biotropica*, 4(2), 117–122.

Gates, G. E. (1972c). *Earthworms: Their Ecology and Taxonomy* (pp. 1–306). Comstock Publishing Associates.

Görres, J. H., Connolly, B. A., & Chang, C.-H. (2014). Earthworm control strategies in forestry ecosystems. *Forest Ecology and Management*, 317, 91–102. https://doi.org/10.1016/j.foreco.2013.11.012

Greiner, H. G., Costello, D. M., & Tiegs, S. D. (2012). Soil biodiversity and agriculture: The role of earthworms in crop growth. *Soil Science Society of America Journal*, 76(4), 1377–1387. https://doi.org/10.2136/sssaj2011.0415

Gundale, M. J., Frelich, M. A., & Reich, P. B. (2005). Susceptibility of a Northern Hardwood forest to exotic earthworm invasion. *Conservation Biology*, 19(4), 1175–1182.

Hale, C. M., Frelich, M. A., & Reich, P. B. (2005). Invasive European earthworm impacts on soil properties and plant communities in northern hardwood forests in Minnesota. *Conservation Biology*, 19(4), 1175–1182.

Hendrix, P. F., & Bohlen, P. J. (2002). Invasive earthworms and their impact on native ecosystems. *Biological Invasions*, 4(1), 87–97.

Howmiller, R. P., & Beeton, A. M. (1971). Influence of organic enrichment on the abundance of aquatic oligochaetes. *Journal of the Water Pollution Control Federation*, 43(3), 494–508.

Howmiller, R. P., & Beeton, R. J. (1971). Earthworms as bioindicators of environmental change. *Ecology*, 52, 139–143.

James, S. W., & Hendrix, P. F. (2004). Earthworm invasions in tropical ecosystems. *Journal of Soil Ecology*, 6(2), 72–78.

Jamieson, B. G. M. (1977a). The indigenous earthworms (Megascolecidae: Oligochaeta) of Tasmania. *Bulletin of the British Museum (Natural History), Zoology*, 32(4), 201–252.

Jamieson, B. G. M. (1977b). Oligochaeta: Their ecology and evolution. *Cambridge University Press*, 1, 27–35.

Jamieson, B. G. M. (2000). *The native earthworms of Australia (Megascolecidae: Megascolecinae)*. Science Publishers: Enfield, NH. (CD-ROM).
Julka, J. M. (1976a). Studies on the earthworms collected during the Daphabum Expedition in Arunachal Pradesh, India. *Records of the Zoological Survey of India*, 69, 229–239.
Julka, J. M. (1976b). Studies on the earthworm fauna of Orissa (India). I. Moniligastridae and Ocnerodrilidae. *Mitteilungen aus dem Zoologischen Museum in Berlin*, 52, 321–329.
Julka, J. M. (1977). Contribution to the knowledge of earthworm fauna (Oligochaeta: Annelida) of Meghalaya. *Newsletter of the Zoological Survey of India*, 3, 398–400.
Julka, J. M. (1978). Studies on the earthworm fauna of Orissa (India). II. Megascolecidae, Octochaetidae and Microchaetidae. *Mitteilungen aus dem Zoologischen Museum in Berlin*, 54, 185–197.
Julka, J. M. (1979). First record of Lumbricus castaneus Savigny from India (Lumbricidae: Oligochaeta). *Journal of the Bombay Natural History Society*, 76, 191–192.
Julka, J. M. (1981). Taxonomic studies on the earthworms collected during the Subansiri Expedition in Arunachal Pradesh, India. *Records of the Zoological Survey of India — Miscellaneous Publication*, Occasional Paper No. 26, 1–53.
Julka, J. M. (1982). Earthworm fauna of the Andaman and Nicobar Islands, India. *Records of the Zoological Survey of India*, 80(1–2), 127–155.
Julka, J. M. (1983). A new genus and species of earthworm (Octochaetidae: Oligochaeta) from South India. *Geobios New Reports*, 2, 48–50.
Julka, J. M., Chandra, M., & Mukherjee, R. N. (1984). Mass migration of earthworms in the Simla Hills. In Director (Ed.), High Altitude Entomology and Wildlife Ecology: Proceedings of the Workshop on High Altitude Entomology and Wildlife Ecology, Solan, Himachal Pradesh (pp. 347–351). *Zoological Survey of India*, Calcutta.
Julka, J. M., & Senapati, B. K. (1986). Earthworms (Oligochaeta: Annelida) of Orissa, India. *Records of the Zoological Survey of India*, Occasional Paper No. 92, 1–49.
Julka, J. M., & Senapati, B. K. (1987). Earthworms (Oligochaeta: Annelida) of Orissa, India. *Records of the Zoological Survey of India*, Occasional Paper No. 92, 1–49. (Reprint/second notation of the 1986 publication.)
Julka, J. M. (1988). The Fauna of India and the Adjacent Countries: Megadrile Oligochaeta (Earthworms). Haplotaxida: Lumbricina: Megascolecoidea: Octochaetidae. *Zoological Survey of India*, Calcutta, xiv + ~400 pp.
Julka, J. M. (1988a). The earthworms of India: Faunal and ecological considerations. In *Fauna of India and the Adjacent Countries: Earthworms (Oligochaeta)* (vol. 1, pp. 98–138). Zoological Survey of India.
Julka, J. M. (1988b). *The Fauna of India and the Adjacent Countries: Megadrile Oligochaeta (Earthworms)—Octochaetidae*. Calcutta: Zoological Survey of India.
Julka, J. M. (1988c). *A Monograph on the Fauna of India: Megadrile Oligochaeta (Earthworms)—Octochaetidae* (p. 400). Calcutta: Zoological Survey of India.
Julka, J. M. (1988d). Taxonomic studies on earthworms in India: A review. *Indian Journal of Soil Science*, 36(4), 441–455.
Julka, J. M. (2014). Invasive earthworms in the Western Ghats region of India. *Soil Biology and Biochemistry*, 71, 76–82.
Julka, J. M., & Paliwal, R. (1993a). Distribution of earthworms in different agro-climatic regions of India. In *Earthworm Resources and Vermiculture*. Calcutta: Zoological Survey of India.
Julka, J. M., & Paliwal, R. (1993b). Distribution of endemic and exotic earthworms in India: A review. *Pedobiologia*, 37(5), 385–399.
Julka, J. M., & Paliwal, R. (1993, 1994). Earthworms of India and their ecological significance. *Environmental Ecology*, 2(2), 97–114.
Julka, J. M., Paliwal, R., & Kathuria, N. (2004). Earthworm checklist of India and their distribution. *Indian Journal of Soil Conservation*, 32(1), 43–59.

Kale, R. D. (1998). Earthworms: nature's gift for utilization of organic wastes. In C. A. Edwards (Ed.), *Earthworm ecology* (pp. 355–376). St. Lucie Press, NY.

Kinberg, J. G. H. (1867). Annulata nova. *Öfversigt af Kongliga Vetenskaps-Akademiens Förhandlingar*, 23, 97–103.

Lavelle, P., Barot, S., & Desjardins, T. (2006). Soil structure and ecosystem functioning: A soil ecology perspective. *Soil Biology and Biochemistry*, 38, 2254–

Lavelle, P., Decaëns, T., Aubert, M., Barot, S., Blouin, M., Bureau, F., Margerie, P., Mora, P., & Rossi, J. P. (2006). Soil invertebrates and ecosystem services. *European Journal of Soil Biology*, 42, S3–S15. https://doi.org/10.1016/j.ejsobi.2006.10.002.

Lee, K. E. (1985). *Earthworms: Their ecology and relationships with soils and land use.* Academic Press.

Lone, S. A., Ahmad, R., & Mir, A. H. (2020). Diversity and distribution of earthworms in the Western Himalayas, India. *Journal of Earthworm Ecology*, 25(2), 129–140.

Lone, S. A., Dar, S. A., & Bhat, S. A. (2022). Taxonomic studies on earthworms of the Kashmir Valley, India. *Biodiversity and Conservation*, 31(3), 741–758.

Loss, S. R., Johnson, S. E., Mabee, T. J., & O'Connell, A. R. (2013). The ecological impacts of invasive earthworms on bird populations. *PLoS ONE*, 8(6), e67788.

Lubbers, M. D., Brussaard, P., Veldkamp, J. F. M., Oenema, E. A. D. M., de Vries, M. J., van der Reest, B. J. M., van der Velden, J. M. T., & van Groenigen, J. A. (2017). Effect of soil acidification on earthworm viability. *Soil Biology & Biochemistry*, 113, 106–114.

Maerz, J. C., Blossey, J. R. L. R. N., & Blossey, L. K. (2009). Declines in woodland salamander abundance associated with non-native earthworm invasions. *Biological Conservation*, 142(3), 615–621.

Mandal, C. K., Reynolds, J. W., Hussain, A., Hasan, M. N., & Ghosh, S. (2023). New distribution records of Octolasion tyrtaeum (Savigny, 1826) and Octolasion cyaneum (Savigny, 1826) (Family: Lumbricidae) earthworm species from Leh-Ladakh Union Territory, India. *Uttar Pradesh Journal of Zoology*, 44(21), 314–320.

Martin, A., and Lavelle, P. (2000). Large-scale effects of earthworms on soil organic matter and nutrient dynamics at a landscape scale over decades. In: Edwards, C. A. (Ed.), *Earthworm Ecology* (2nd ed., pp. 145–160). Boca Raton, Florida: CRC Press.

Martins, M., Ferreira, M. A., & Silva, A. (2008). Earthworms as bioindicators in agroecosystems. *Pedobiologia*, 51, 53–61.

Martins, R. T., Martins, S. P., da Silveira, M. S. C., de Sousa, F. A., & de Melo, A. M. L. S. T. (2008). Earthworms as bioindicators of soil quality in Brazilian agricultural systems. *Pesquisa Agropecuária Brasileira*, 43(1), 105–112.

Michaelsen, W. (1900). *Das Tierreich, Volume 10: Oligochaeta.* Berlin: R. Friedländer und Sohn.

Michaelsen, W. (1903). The discovery of peregrine earthworm species in the early 20th century. *Zoologica*, 9(5), 117–121.

Michaelsen, W. (1907a). The earthworms of India. *Zoologische Jahrbücher, Abteilung für Systematik, Geographie und Biologie der Tiere*, 25, 555–569.

Michaelsen, W. (1907b). Oligochaeta. In *Das Tierreich.* Berlin: Friedländer & Sohn.

Michaelsen, W. (1910a). Die Oligochaeten des Indischen Subkontinents. *Abhandlungen der Senckenbergischen Naturforschenden Gesellschaft*, 32, 1–117.

Michaelsen, W. (1910b). The earthworms of India. *Zoologische Jahrbücher, Abteilung für Systematik, Geographie und Biologie der Tiere*, 27, 312–321.

Mısırlıoğlu, M., Küpeli, S., & Arslan, M. (2023). Ecological functions and biodiversity of earthworms in diverse soil habitats. *Journal of Soil Science and Environmental Management*, 14, 52–65.

Mısırlıoğlu, M., Reynolds, J. W., Stojanović, M., Trakić, T., Sekulić, J., James, S. W., Csuzdi, C., Decaëns, T., Lapied, E., Phillips, H. R. P., Cameron, E., & Brown, G. G. (2023). Earthworms (Clitellata, Megadrili) of the world: An updated checklist of valid species and families, with notes on their distribution. *Zootaxa*, 5255(1), 417–438.

Naik, A., Mahata, A., & Palita, S. K. (2024). Studies on earthworm diversity with respect to soil properties in different land use systems in Koraput region of the Eastern Ghats, India. *Biodiversity and Conservation*, 33(2), 3355–3376.

Narayanan, S. P., Harit, A. K., Singh, M., Bhattacharyya, S., Dutta, A., & Thomas, A. P. (2023a). A contribution to the earthworm (Annelida, Clitellata, Megadrili) fauna of the central Indian state Chhattisgarh, India. *Opuscula Zoologica*, 54(2), 185–194.

Narayanan, S. P., Julka, J. M., & Paliwal, R. (2020a). Earthworm diversity in the Western Ghats, India: A review. *Zootaxa*, 4755(1), 1–68.

Narayanan, S. P., Kumari, S., Kurien, V. T., Thomas, A. P., Paliwal, R., & Julka, J. M. (2021a). A comprehensive checklist of the earthworms (Annelida: Clitellata: Megadrili) of Sri Lanka, a component of the Western Ghats – Sri Lanka biodiversity hotspot. *Travaux du Muséum National d'Histoire Naturelle "Grigore Antipa"*, 64(1), 7–36. https://doi.org/10.3897/travaux.64.e56877

Narayanan, S. P., Lone, M. Y., Sathrumithra, S., Anuja, R., & Thomas, A. P. (2023b). Earthworms (Clitellata: Moniligastrida, Crassiclitellata) of India: Distribution and Status. *Opuscula Zoologica*, 54(2), 185–194.

Narayanan, S. P., Paliwal, R., Kumari, S., Ahmed, S., Thomas, A. P., & J. M. Julka. (2020b). Annelida: Oligochaeta, 87–102 pp.

Narayanan, S. P., Sudhakaran, A. R., & Thomas, P. (2020c). Taxonomy and distribution of earthworms in the Western Ghats of India. *Journal of Earthworm Ecology*, 16, 42–60.

Narayanan, S. P., Sudhakaran, A. R., & Thomas, P. (2021b). Ecological patterns of earthworm communities in the Western Ghats. *Journal of Soil Science and Environmental Management*, 32, 1–18.

Narayanan, S. P., Sudhakaran, A. R., & Thomas, P. (2022b). Earthworm diversity in India: Insights into species ecology. *Ecological Indicators*, 98, 29–41.

Narayanan, S. Prasanth, Jose, Divya, Kurien, Vijo Thomas & Thomas, A. P. (2024a). On three new species of the earthworm genus Drawida Michaelsen, 1900 (Clitellata: Moniligastridae) from south-western India. *Zootaxa*, 5541(4), 563–581.

Nouri-Aiin, M., & Görres, J. H. (2021). Laboratory evaluation of entomopathogenic fungus *Beauveria bassiana* for biological control of invasive earthworm. *Pheretima hawayana*. *PeerJ*, 9, e11101. https://doi.org/10.7717/peerj.11101

Ozdemir, S., D. A. Demiray, A. O. C. Ercan, & M. S. Cicek. (2011). Use of Earthworms as Bioindicators in the Assessment of Agricultural Soil Pollution. *Journal of Environmental Protection*, 2(7), 882–889. https://doi.org/10.4236/jep.2011.27101

Paliwal, R., & Julka, J. M. (2005). A comprehensive checklist of Indian earthworms. *Zoological Survey of India*, 47, 102–113.

Paliwal, R., & Julka, J. M. (2005). Checklist of earthworms of western Himalaya, India. *Zoos' Print Journal*, 20(9), 1972–1976.

Pelosi, C., & Röembke, J. (2016a). Bioindicator species in the evaluation of soil and ecosystem health. *Environmental Toxicology and Chemistry*, 35, 1079–1086.

Pelosi, C., & Röembke, J. (2016b). Earthworms as bioindicators in agricultural soils of Europe. In *Soil Biodiversity in Agricultural Systems*. CRC Press.

Razafindrakoto, E. A., Ravelomanana, L., & Rasoanivo, A. (2010). Biodiversity and ecological studies of earthworms in tropical Madagascar. *Soil Biology and Biochemistry*, 42, 1164–1172.

Razafindrakoto, M., Csuzdi, C., Rakotofiringa, S., & Blanchart, E. (2010). Earthworm diversity in Madagascar. *Applied Soil Ecology*, 45(3), 201–208.

Reynolds, J. W., & Wetzel, M. A. (2024a). Earthworms of North America: A review of their ecology and role in soil health. *Soil Biology and Biochemistry*, 132, 185–196.

Reynolds, J. W., & Wetzel, M. A. (2024b). Ecological impacts of invasive earthworms on soil biota. *Soil Science*, 180(4), 349–359.

Reynolds, J. W., & Wetzel, M. J. (2024). *Nomenclatura Oligochaetologica – A catalogue of names, descriptions and type specimens.* Editio Secunda. www.inhs.illinois.edu

Rodingpuia, M., Lalthanzara, H., Lalfakzuala, R., & Lalmuanliah, C. (2024). Earthworm biodiversity and its role in soil health. *Asian Journal of Soil Science*, 41(5), 263–271.

Römbke, J., Aira, M., & Domínguez, J. (2013). A critical review of the use of earthworms in the risk assessment of chemicals in soil. In: *Progress in Soil Biology and Ecology* (pp. 215–234). Boca Raton, Florida: CRC Press/Taylor & Francis Group.

Rouse, G. W., & Pleijel, F. (2007). Annelid Phylogeny and Systematics. In G. W. Rouse, F. Pleijel, & E. Tilic (Eds.), *Reproductive Biology and Phylogeny of Annelida* (pp. 42–60). CRC Press.

Satchell, J. E. (1983). Methods of sampling earthworm populations. In: *Earthworm Ecology: From Darwin to Vermiculture* (ed. J. E. Satchell), pp. 8–13. Chapman & Hall, London.

Sautter, K. D., Brown, G. G., James, S. W., Pasini, A., Nunes, D. H., & Benito, N. P. (2006a). Present knowledge on the diversity of earthworms in Brazil and their potential use. *European Journal of Soil Biology*, 42(Suppl 1), S145–S149. https://doi.org/10.1016/j.ejsobi.2006.07.042

Sautter, G., Lopes, M., & Guérin, P. (2006b). Earthworm species richness in agricultural ecosystems. *Environmental Biology of Fishes*, 77, 57–66.

Senapati, B. K. (1992). Vermicomposting: an option for recycling of cellulosic wastes of India. In: *New Trends in Biotechnology* (eds. N. S. Subba Rao, C. Balagopalan and S. V. Ramakrishna), pp. 347–358. Oxford and IBH Pub., New Delhi.

Senapati, B. K., Dash, M. C., & Paliwal, R. (2022). Earthworm ecology in India. In N. Karmegam (Ed.), *Earthworm Ecology and Biogeography in India* (pp. 1–20). Singapore: Springer.

Stephenson, J. (1914). Littoral Oligochaeta from the Chilka Lake on the East Coast of India. *Records of the Indian Museum*, 10, 225–262.

Stephenson, J. (1914). On a collection of Oligochaeta mainly from Northern India. *Records of the Indian Museum*, 10, 321–365.

Stephenson, J. (1915). On some Indian Oligochaeta, mainly from Southern India and Ceylon. *Memoirs of the Indian Museum*, 6, 35–108.

Stephenson, J. (1916). On a collection of Oligochaeta belonging to the Indian Museum. *Records of the Indian Museum*, 12, 299–354.

Stephenson, J. (1917). On a collection of Oligochaeta from various parts of India and further India. *Records of the Indian Museum*, 13, 353–416.

Stephenson, J. (1920). On a collection of Oligochaeta from the lesser-known parts of India and from eastern Persia. *Memoirs of the Indian Museum*, 7, 191–261.

Stephenson, J. (1921). Contribution to the morphology, classification, and zoogeography of Indian Oligochaeta. I. The affinities and systematic position of the genus Eudichogaster Michaelsen, and some related questions. II. On polyphyly in the Oligochaeta. III. Some general considerations in the geographical distribution of Indian Oligochaeta. *Proceedings of the Zoological Society of London*, 1921, 103–142.

Stephenson, J. (1922a). On some earthworms from Kashmir, Bombay, and other parts of India. *Records of the Indian Museum*, 24, 427–443.

Stephenson, J. (1922b). On some earthworms from India and Palestine belonging to the British Museum. *Annals and Magazine of Natural History, Series 9*, 9(52), 129–136.

Stephenson, J. (1923). Oligochaeta. In *The Fauna of British India, including Ceylon and Burma*. Taylor & Francis, London, xxiv + 518 pp.

Stephenson, J. (1923). *The Fauna of British India, Including Ceylon and Burma: Oligochaeta* (p. 518). London: Taylor and Francis.

Stephenson, J. (1930). *The Oligochaeta*. Oxford: Clarendon Press.

Suárez, M. L., Lozano, Y. M., & Hättenschwiler, S. (2006). The effect of earthworms on forest soil erosion rates. *Journal of Soil and Water Conservation*, 61(2), 122–129.

Timm, L. M. (2012). *Earthworms of the World: Ecology, Behavior, and Distribution*. Springer Science & Business Media.

Tiwari, K., Rodrigues, S., Mehta, P., & Iyer, A. (2024). Exploring the ecology of native and invasive earthworms in the Western Ghats. *Journal of Soil Biology*, 29(3), 243–251.

Tiwari, S., Yadav, P., & Yadav, S. (2021). New taxa of earthworms in the Northeast region of India. *Ecology of Earthworm Fauna*, 45, 78–89.

Velki, M., & Ecimović, V. (2017). Use of earthworms as bioindicators in contaminated soils. *Environmental Pollution*, 223, 249–260.

Verma, R., & Yadav, P. (2010). Earthworm fauna of Uttar Pradesh. *Journal of Environmental Biology*, 31, 343–350.

Weber, A. (2007). Earthworms as bioindicators in ecosystems affected by human activities. *Environmental Pollution*, 148, 45–56.

Yadav, P. (2011). Ecological studies of earthworms from Rajasthan. *Indian Journal of Ecology*, 38, 43–50.

Yadav, P. (2016). Taxonomic review of earthworms of Rajasthan, India. *Journal of Soil Science and Environmental Management*, 14, 101–109.

Zhang, Z., P. F. Hendrix, and P. Lavelle. (2019). Biological control options for managing invasive earthworm species. *Ecological Management*, 13(4), 350–358.

Zhang, Z., Chen, X., Li, J., Lv, X., Wu, H., & Hu, F. (2020). The role of earthworm bioturbation in increasing CO_2 emissions. *Global Change Biology*, 26(8), 2257–2266.

3 Habitat and Ecology of Terrestrial Earthworms

3.1 INTRODUCTION

The physical environment where an organism or population resides includes both biotic and abiotic components, and organisms are known to interact with their environment. In India, earthworms have been studied extensively across various habitats, revealing their crucial roles in soil ecology and ecosystem sustainability.

Studies by Julka (1988a,b) and later by Julka and Paliwal (2005) reveal that earthworms thrive in moist, nutrient-rich soils, particularly in agricultural fields and gardens, where their activities significantly enhance soil fertility through decomposition and nutrient cycling. Julka in 1983 observed that these organisms are often found under rocks and logs, taking advantage of moisture and protection.

India, as a country rich in its biodiversity, has a unique collection of earthworms. Stephenson (1923) noted unique adaptations in tropical forests, where some earthworm species migrate to live in epiphytic plants on trees, revealing diversity of habitats. This diversity of habitats, ranging from puddles to forests, underscores the adaptability of earthworms and their importance in India's varied landscapes.

There are more than 450 species of earthworms reported from different families including Acanthodrilidae, Glossoscolecidae, Lumbricidae, Almidae, Eudrilidae, Moniligastridae, Ocnerodrilidae, Octochaetidae, Rhinodrilidae, Megascolecidae, under which genus Perionyx is abundantly distributed in India. With advances in molecular techniques of barcoding, several new records and species are being added to the existing list of earthworms of India (Ghosh, 2022; Ghosh et al., 2022b; Mandal et al., 2021; Hasan et al., 2019; Ghosh et al., 2018, 2023; Mandal et al., 2023) and across the globe.

Lately computational tools are being used to understand the distribution of earthworms (Ghosh et al., 2022a). More and more studies using diverse tools are being used to understand the biochemistry and molecular biology. (Ghosh et al., 2020; Ghosh, 2020a, 2020b; Ghosh and Das, 2024; Ghosh et al., OM536159, OM536160, OM536161).

These earthworms mostly are found to reside in soil and in fresh water. The soil dwellers reveal ecotypes (Figure 3.1) of (1) epi-endogeic ecotypes like *Dendrobaena veneta* (*D. veneta*) and *Eisenia fetida* (*E. fetida*), found in warm and moist environments, under litter, rapidly consuming and disposing waste and soil contaminants; (2) epigeic ecotypes, predominantly including *Lumbricus castaneus (L. castaneus), Lumbricus festivus (L. festivus), Lumbricus rubellus (L. rubellus), Dendrobaena octaedra (D. octaedra), Dendrodrilus rubidus (D. rubidus), Eiseniella tetraedra (E. tetraedra), Satchellius mammalis (S mammalis), Dendrobaena attemsi (D. attemsi),* and *Heliodrilus oculatus (H. oculatus),* living on the soil surface, thriving on litter

DOI: 10.1201/9781003633600-3

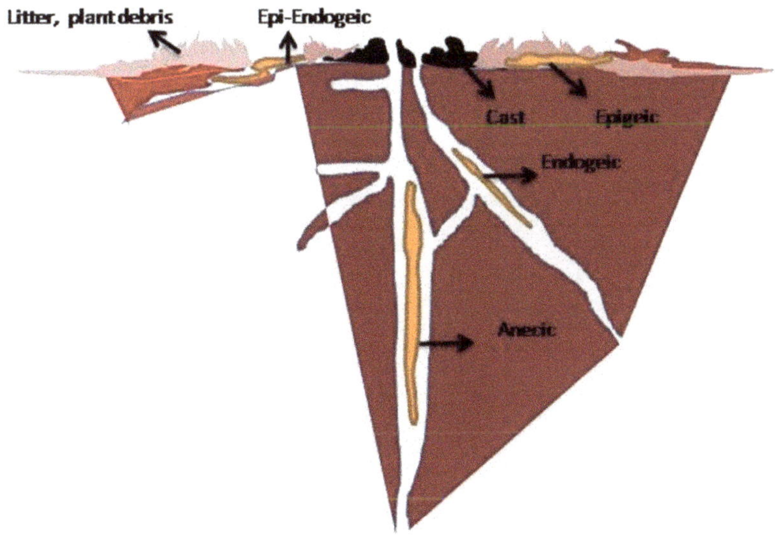

FIGURE 3.1 Ecotypes of soil-dwelling earthworms.

(Image adapted from Ghosh, 2022).

and organic matter; (3) endogeic ecotypes, including *Octolasion cyaneum (O. cyaneum), Octolasion lacteum (O. lacteum), Allolobophora chlorotica (A. chlorotica), Apporectodea caliginosa (A. caliginosa), Apporectodea icterica (A. icterica), Apporectodea rosea (A. rosea),* and *Muchieona muldali (M. muldali)*, found in soil within 20 cm depth, making horizontal burrows, feeding on soil, making semi-permanent burrows horizontal in direction; and (4) aneic ecotypes, which are subsoil dwellers, found at 3 m below the top soil surface, predominantly including *Lumbricus terristris (L. terristris) Apporectodea longa (A. longa), Apporectodea nocturna (A. nocturna),* and *Lumbricus friendi (L. friendi)*, making permanent vertical burrows in the soil, feeding on leaves, and releasing casts at burrow entrances. Aquatic Oligochaetes residing on moist soil, around freshwater bodies, including ponds and lakes, also reveal benthic lifeforms, thriving on organic matter in ponds, stream bottoms, sediments, and upper layers of the benthic surface, with a parasitic existence. Species under the *Glyphidrilus* Horst genus reported from rivers in Thailand reveal semi-aquatic freshwater life (Ghosh, 2022).

Studies focused on the ecology of Indian earthworms highlight their significant contribution to agricultural productivity by improving soil structure, aeration, and water infiltration through their burrowing activities. Such benefits have spurred growing interest in vermiculture and sustainable agriculture, where earthworms are harnessed for composting and organic farming (Kale, 1998).

The adaptability of Indian earthworms to tropical and subtropical conditions further influences their distribution and ecological roles. Their presence is also associated with increased microbial activity and nutrient availability in the soil, which aids plant growth and supports biodiversity conservation efforts.

In India, researchers continue to explore earthworm ecology. There is increasing recognition of their potential in enhancing sustainable agriculture, improving soil health, and contributing to environmental resilience, which ultimately supports both agricultural output and ecological stability (Figures 3.2 and 3.3).

FIGURE 3.2 Life cycle of *E. lucens*: (a) adults (0.9–2.2 g); (b) cocoons (approx. 50 mg); (c) hatchlings (5–35 mg).

(Image adapted from Kostecka et al., 2020. Under Open Access Creative Commons Attribution 4.0 International License).

FIGURE 3.3 Terrestrial earthworms collected from habitat, India.

3.2 TERRESTRIAL HABITATS

Terrestrial habitats, ranging from vast forests and arid deserts to open grasslands and wetlands, provide essential environments for a myriad of species. Earthworms play a significant role in maintaining soil health and supporting ecological balance across diverse environments.

The interactions between abiotic factors, including temperature, moisture, and soil composition and biotic factors such as plant roots, microorganisms, and other soil fauna, shape the distribution and ecological functions of earthworms within these habitats. In forests, earthworms are crucial to the decomposition of organic material, recycling nutrients and enhancing soil structure.

These environments, often characterized by high humidity and constant temperatures, offer ideal conditions for earthworms, particularly in the tropical forests of India. Julka (1988a,b) and Bhardwaj et al. (2015) observed that earthworms thrive in the organic-rich soil layers created by fallen leaves and decaying wood.

Julka (1983a,b) noted that the diversity of earthworm species in Indian forests is associated with soil moisture and the availability of organic material. These organisms contribute to nutrient recycling by breaking down plant matter and enriching the soil with essential nutrients like nitrogen (N) and phosphorus (P).

The Western Ghats in India is a hotspot for earthworm species that facilitate nutrient recycling, thus supporting the diverse plant and animal life in these ecosystems (Kale, 1998).

Desert ecosystems, despite their extreme temperatures and water scarcity, also support certain earthworm species that are specially adapted to survive under these harsh conditions.

Noy-Meir (1973) observed that species of earthworms surviving in the desert have evolved mechanisms to conserve water and withstand high temperatures.

In India, the Thar Desert is home to earthworm species that play an important role in breaking down sparse organic material and enhancing soil structure. These earthworms contribute to soil aeration and improve water retention, which is critical in desert habitats.

The adaptability of earthworms to such environments highlights their ecological importance, even in the most arid conditions (Singh & Gupta, 2010).

In the grassland ecosystem of the Indian savannas, earthworms contribute significantly to soil health by decomposing plant matter, thus promoting the cycling of nutrients essential for plant growth.

These habitats are often dominated by grasses and small shrubs, and earthworms play a role in maintaining the productivity of these ecosystems by enriching the soil. Singh (2012) highlighted that in grassland ecosystems, earthworms improve soil structure through their burrowing activities, which enhance water infiltration and root development.

Earthworms also contribute to the growth of microbial communities that play an important role in the decomposition of plant material, contributing to nutrient recycling in these expansive habitats. Wetlands, including marshes, swamps, and floodplains, are among the most biodiverse terrestrial habitats, providing breeding grounds for amphibians, waterfowl, and various aquatic plants.

Mitsch and Gosselink (2007) noted that wetland earthworms enhance soil fertility, improving the growth of vegetation that sustains a diverse array of organisms. Moreover, their burrowing activities increase water retention and contribute to flood regulation by enhancing the permeability of wetland soils.

Earthworms in wetland habitats of Indian wetlands, including Sundarbans and Bharatpur, have been reported to play a role in breaking down organic material and recycling nutrients that support aquatic and terrestrial plant species.

In addition to larger terrestrial ecosystems, earthworms can be found in microhabitats within these broader environments, including niches of leaf litter, tree bark, and under rocks, which provide specialized conditions for earthworms and other soil organisms. These microhabitats play an essential role in maintaining the ecological balance of broader terrestrial environments by supporting species that might not thrive in the larger, more competitive habitat.

Schowalter (2000) observed that leaf litter in forests is home to decomposers like earthworms, fungi, and various arthropods, which break down organic matter and recycle nutrients back into the soil. Bergman (2002) observed tree bark as the preferred habitat for small earthworm species thriving on decomposing organic matter and interacting with the microbial communities on tree surfaces.

Across all these terrestrial habitats, earthworms contribute to soil health by improving soil structure, enhancing water infiltration, and promoting nutrient recycling. The burrowing activity of earthworms creates channels that allow penetration of water into the soil, improving soil aeration and preventing erosion. The organic matter they ingest and digest is broken down into humus, leading to enrichment of soil and providing essential nutrients for plants. Earthworms also facilitate the growth of beneficial soil microorganisms, creating a healthy soil ecosystem that supports plant and animal life.

3.3 ABIOTIC FACTORS

Temperature, water, light, soil composition, and air are major abiotic factors that significantly influence the distribution, survival, and behavior of organisms in terrestrial ecosystems, including earthworms. These factors shape the environmental conditions in which earthworms, as well as other organisms, thrive or struggle to survive. These abiotic factors play an important role for terrestrial earthworms, directly affecting their metabolism, reproduction, and ecological functions.

1. Temperature has a profound influence on the metabolism and reproduction of earthworms, as it does for many other organisms. Earthworms are ectothermic, meaning their body temperature is regulated by the surrounding environment, which directly affects their metabolic rate.

   Hochachka and Somero (2002) observed that higher temperatures accelerate metabolic processes in ectotherms, including earthworms, leading to higher rates of digestion and nutrient cycling. However, extreme temperatures have proven detrimental effects. Earthworms are generally adapted to specific temperature ranges, and when exposed to temperatures outside their tolerance limits, they may exhibit reduced activity, slowed growth, or even death.

   In India, studies by Julka (1983a,b) and Kale (1998) have demonstrated that earthworms thrive in temperate regions with moderate temperatures. In the tropical climates of the Western Ghats, earthworms are adapted to survive in consistently warm conditions. However, in colder regions of India, earthworm activity tends to decrease, especially during the winter months, as the soil temperature drops.

   Earthworms found in the Himalayan foothills (Singh, 2012) experience seasonal variations in activity, where they become dormant during the colder months and resume their ecological functions as temperatures rise.
2. Water availability is crucial for earthworm survival, as these organisms rely on moisture to maintain proper hydration and facilitate respiration through their skin. Earthworms are typically found in moist environments where soil moisture levels are high. According to Noy-Meir (1973), in dry conditions, earthworms are often absent or inactive, and their populations in arid regions like the Thar Desert are limited. Conversely, in more humid regions like the Sundarbans or Kerala's wetlands, earthworms are abundant due to the consistently high moisture content in the soil, which supports their metabolic processes and reproductive cycles.
3. Water salinity also impacts earthworm populations, especially in coastal areas, as observed by Gibbs (2001). While in the brackish water environments of the Indian coast, with fluctuating soil salinity, only certain earthworm species tolerant of salt can thrive, Kale and Bano (2003a,b,c) highlighted that Indian earthworm species in coastal ecosystems exhibit remarkable adaptability to changing salinity conditions, which may influence their habitat selection and ecological roles.

4. Light availability influences earthworm behavior indirectly by impacting their activity patterns. Earthworms are nocturnal and prefer to stay in the dark, moist environments of soil to avoid desiccation and predation. Diurnal patterns of activity, where earthworms are more active at night, are primarily dictated by light cycles (Daan et al., 2001).
   In tropical forests, with diffused light seeping in through dense canopies, earthworms thrive in the moist, shaded undergrowth (Schowalter, 2000), while Kale (1998) suggested that in temperate forests, with less intensity of light penetration, earthworms prefer the deeper layers of the soil in order to maintain moisture levels and avoid temperature extremes and tend to avoid surface exposure. In agricultural soils, light indirectly influences earthworm populations by determining the presence and density of plant cover.
5. The extent of photosynthesis in plants can affect the amount of organic matter available to earthworms, as observed by studies by Field (1998). Greater plant growth provides more organic material in the form of leaves and roots, which supports earthworm populations. Wetzel (2001) highlighted that light availability affects the photosynthetic organisms in aquatic ecosystems, which in turn influences the terrestrial nutrient inputs in nearby ecosystems, thereby having an impact on earthworm habitats.
6. The soil composition, including its nutrient content, texture, and pH, is crucial for earthworm distribution and activity. Earthworms prefer loamy soils rich in organic matter, where they can easily burrow and obtain food in the form of decomposing plant material.
   According to Jones (2004), soil texture influences the ease with which earthworms can move and create burrows.
7. Soil pH is another important factor, as earthworms generally prefer neutral to slightly alkaline soils, with pH ranges between 6 and 8. Acidic soils with a pH lower than 5.5 can be hostile to earthworm populations, as observed by Rennie et al. (1996). In India, the acidic soils of northeast India can limit the abundance of earthworms, while more neutral soils in regions like the Deccan Plateau and Himalayan foothills support a diverse array of earthworm species (Bhardwaj et al., 2015).
8. Oxygen is essential for the respiration of earthworms, as they rely on oxygen diffusing through their skin. Earthworms in well-oxygenated soils are more active and able to perform their ecological functions more effectively.
9. High carbon dioxide levels, on the other hand, can indicate poor soil aeration and hinder earthworm activity. Schlesinger (1997) noted that excessive CO_2 in the soil can occur in compacted or over-watered soils, which can cause earthworms to become stressed or dormant.
10. Humidity plays a critical role in the survival of earthworms, as they are highly susceptible to desiccation. In humid environments such as wetlands and tropical forests, earthworms are more active and have a higher survival rate. However, in arid regions or during dry periods, earthworms tend to enter a state of dormancy, where they seal themselves within their burrows to conserve moisture (McDowell, 2011).

3.4 BIOTIC FACTORS

Biotic factors are the living components of an ecosystem that influence interactions between organisms and shape the functioning of ecosystems. These factors include producers; consumers; decomposers; and various interspecific interactions such as predation, competition, and symbiosis. In the context of terrestrial earthworms, these biotic factors play a crucial role in their survival, behavior, and ecological functions.

Earthworms are critical in nutrient cycling, soil formation, and ecosystem dynamics, influencing and being influenced by various biotic factors.

In terrestrial ecosystems, plants are the primary producers, providing food for herbivores and ultimately sustaining entire food webs. Earthworms are intricately linked to plant productivity and the organic matter produced by plants. Earthworms play a key role in soil health by consuming plant material and decomposing it, enriching the soil with nutrients. They are particularly active in environments with abundant plant matter, such as tropical forests and agricultural lands in India. In Indian ecosystems, rice and wheat are dominant agricultural crops that support primary production, while mangrove plants in coastal areas also contribute significantly to the primary production of these ecosystems.

Experiments have suggested that earthworms can be grown in laboratories just like in field conditions. The adult *Eisenia lucens* (Waga 1857) life cycle in the wild was observed and revealed similarity to that of earthworms grown under artificial laboratory conditions with an artificial diet of manure (Kostecka et al., 2020). Studies by Verma and Soni (2012) revealed that earthworms in Indian tropical forests, in soils enriched with organic material from plants, play a critical role in nutrient cycling. Similarly, in wetlands such as the Sundarbans, where primary production is high due to the abundance of aquatic plants, earthworms help in decomposing plant material and enhancing soil fertility (Basu & Ghosh, 2003).

Consumers are organisms that obtain their food by consuming other organisms, including herbivores, carnivores, omnivores, and detritivores. Earthworms are pri marily detritivores, meaning they feed on decomposing organic matter, including dead plant material and animal remains, thereby recycling nutrients back into the soil, which are then available to producers (plants). In Indian ecosystems, herbivores such as the nilgai and blackbuck graze on grasses and are known to indirectly influence earthworm populations by affecting the amount of organic material available for decomposition (Chakraborty et al., 2010).

Carnivores like the Indian leopard, preying on herbivores, also have an indirect effect on earthworms by shaping the herbivore population dynamics and, in turn, the organic matter input into the soil. Earthworms interact with a range of detritivores and decomposers such as fungi and microorganisms. In particular, the fungus genus *Mycorrhiza* forms mutualistic relationships with plants, enhancing nutrient absorption, which in turn affects the organic material available for earthworms (Pande & Reddy, 2010a,b). Earthworms help break down this organic matter, returning important nutrients such as nitrogen and phosphorus to the soil. This nutrient cycling is essential for maintaining soil fertility in agricultural systems, where earthworms enhance plant growth by enriching the soil.

Decomposers, including fungi, bacteria, and other microorganisms, are responsible for breaking down dead organic matter into simpler substances that are then recycled back into the environment. Earthworms are key decomposers in terrestrial ecosystems, as they ingest organic matter and excrete it in a digested form that is more accessible to microorganisms. Research by Kale and Bano (2003a,b,c) emphasizes the importance of earthworms in enhancing soil structure and fertility through the breakdown of organic material. In forests, where large amounts of leaf litter accumulate, earthworms help decompose the material, facilitating the recycling of nutrients and supporting the food web (Singh, Kaur, & Verma, 2006). In the Himalayan foothills and temperate forests of India, where leaf litter is abundant, earthworms contribute significantly to soil processes by consuming organic matter and enriching it with vital nutrients for plant growth.

Studies by Julka (1983a,b) show that these decomposers are especially abundant in moist and nutrient-rich soils, where they contribute to the soil's organic carbon content and nutrient cycling.

Predation is a relationship in which one organism (the predator) hunts and consumes another (the prey). While earthworms are not typically predators, they act as prey for a variety of predators, including birds, amphibians, reptiles, and small mammals. Birds such as the Indian robin and kingfisher often feed on earthworms, while mammals like the Indian mole and wild boar also prey on them. In ecosystems where these predators are abundant, earthworm populations can be regulated, which in turn affects soil structure and nutrient cycling. In Indian ecosystems, the Bengal tiger and Indian leopard indirectly affect the availability of earthworms by regulating populations of herbivores like deer that influence plant material available for decomposition.

Predation helps to maintain the balance of ecosystems by ensuring that no one species overexploits a particular resource. Competition occurs when organisms vie for the same resources, such as food, water, or territory. In the case of earthworms, competition can arise with other detritivores, such as beetles or other species of earthworms. In forests or agricultural lands, earthworms often share resources with other detritivores that feed on decaying organic matter.

Competition can influence earthworm populations and their activity, with larger, more aggressive species sometimes outcompeting smaller ones. In tropical ecosystems with high biodiversity, like India's Western Ghats, earthworms may compete for food resources with other soil organisms such as fungi and microorganisms (Srinivasan, & Chandran, 2002). The outcome of this competition can shape the distribution and abundance of earthworms and influence the rates of nutrient cycling.

Symbiosis refers to close interactions between different species, which can be mutually beneficial (mutualism), neutral (commensalism), or harmful (parasitism). Earthworms engage in various symbiotic relationships that benefit both the earthworms and other species. A well-known example is the mutualistic relationship between earthworms and fungi in the Mycorrhizal association, where the fungi enhance nutrient uptake for plants, and earthworms help decompose plant material, recycling nutrients back to the soil (Pande & Reddy, 2010a, 2010b).

Additionally, earthworms have mutualistic relationships with microorganisms in the soil. These microorganisms help break down organic matter that earthworms consume, while the earthworms provide a habitat and food for the microorganisms. This interaction is vital for nutrient cycling and soil health.

3.5 THE ROLE OF EARTHWORMS IN ECOLOGICAL NICHES

An ecological niche refers to the specific role an organism plays within its environment, including its interactions with other species, its habitat, and its role in the flow of energy and nutrients. The ecological niche concept is fundamental for understanding the interactions among species and how they coexist, especially in ecosystems where resources are limited or competition is high.

In the case of terrestrial earthworms, their ecological niche involves their habitat; their role in decomposition; and their interactions with other organisms, such as plants, microorganisms, and other detritivores. The study of niche differentiation and resource partitioning helps to understand how earthworms and other species reduce competition and maintain ecosystem stability. Earthworms play a critical role in terrestrial ecosystems, especially in soil health and nutrient cycling. Their ecological niche involves consuming organic matter, such as dead plant material, and breaking it down into simpler forms that are more accessible to other organisms like plants and microorganisms. This process is integral to maintaining soil fertility, which, in turn, supports plant productivity and sustains higher trophic levels.

In the Indian subcontinent, earthworms occupy diverse niches depending on their habitat. For instance, in agricultural ecosystems, earthworms often live in the topsoil, where they consume decaying plant material and organic fertilizer. In forests, they are more deeply embedded in the forest floor, feeding on leaf litter, roots, and decaying organic matter, which helps recycle nutrients and improve soil structure.

Research by Julka (1983a,b) has shown that different species of earthworms are adapted to specific environmental conditions in India, with some species thriving in moist, nutrient-rich soils and others in more arid or disturbed environments.

This ecological flexibility highlights the importance of earthworms in various ecosystems, ranging from agricultural lands to temperate forests and wetlands. In Indian tropical forests, where biodiversity is high, earthworms play a crucial role in nutrient cycling.

Ranganathan and Sreenivasan (1999a,b) suggest that the niche differentiation of earthworms, based on their feeding behavior and habitat preference, helps reduce competition with other detritivores, such as termites and beetles, that also play a role in decomposition. Resource partitioning refers to the division of resources by similar species to minimize competition. It can take the form of spatial, temporal, or trophic segregation. In terrestrial ecosystems, earthworms exhibit resource partitioning in various ways, depending on their ecological niche.

In Indian forests and agricultural systems, earthworms can partition resources spatially by occupying different soil layers. Some earthworm species, such as *Lumbricus terrestris*, are anecic (surface-dwelling) species, meaning they feed on surface organic material and create deep burrows.

Other species, such as endogeic (burrow-dwelling) earthworms, feed on decaying plant material deeper in the soil. This spatial partitioning allows multiple earthworm species to coexist within the same ecosystem without directly competing for the same resources (Kale & Bano, 2003a,b,c).

In agricultural ecosystems, earthworms also exhibit spatial resource partitioning by feeding on different layers of soil and organic matter. This behavior is particularly important in maintaining soil structure and fertility.

As Singh, Kaur, & Verma (2006) have shown, earthworms in Indian agricultural lands help aerate the soil, improve water infiltration, and break down organic material, thereby enhancing soil health. Temporal resource partitioning is also a significant mechanism that helps reduce competition among species.

While this is more commonly observed in herbivores or carnivores, some studies have suggested that earthworms may show some form of temporal partitioning, particularly with regard to feeding behavior. In some agricultural systems, earthworms feed more actively during the cooler hours of the day or during the monsoon season, when organic material is more abundant. This helps avoid direct competition with other detritivores, such as fungi and bacteria, which might decompose organic matter at different rates depending on environmental conditions (Pande & Reddy, 2010a,b).

Trophic partitioning refers to the different feeding strategies that species use to reduce competition. While most earthworms are detritivores, they may specialize in different types of organic material. Some species primarily feed on leaf litter, while others specialize in decaying roots or animal matter. This trophic partitioning is essential for the ecological role of earthworms in nutrient cycling, as it ensures that they do not directly compete with other decomposers for the same resources.

In Indian forests, species like *Eisenia fetida* and *Aporrectodea longa* specialize in different types of organic matter. Studies by Reddy et al. (2000) show that these species coexist by partitioning the resources available to them based on their feeding preferences and ecological roles. Resource partitioning is observed across various species in Indian ecosystems, and it allows organisms with similar resource needs to coexist. In the Sundarbans, for example, mangrove crabs exhibit resource partitioning by feeding on different types of detritus and occupying specific zones within the mangrove habitat (Chakraborty & Saha, 2011a,b).

Similarly, in the Indian Himalayas, herbivorous animals such as the Himalayan tahr and bharal (blue sheep) partition resources by grazing on different types of vegetation, with the tahr feeding on shrubs and the bharal on grasses (Sati & Awasthi, 2006). In Indian grasslands, temporal resource partitioning is observed between herbivores like the Indian gazelle and blackbuck. The blackbuck primarily feeds during the cooler parts of the day, while the gazelle feeds in the early morning and late evening, thus avoiding direct competition for food (Jhala & Nayak, 2008).

Similarly, different bird species in wetland ecosystems such as Keoladeo National Park partition resources by feeding on specific types of prey and occupying specific zones in the wetland (Kumar & Singh, 2005). In Indian agricultural ecosystems, ecological niche differentiation is also evident. Earthworms play an essential role in soil fertility, and their activity can be influenced by the types of crops grown and the availability of organic material. Indian honeybees and other pollinators exhibit a unique ecological niche by facilitating plant reproduction, which indirectly benefits earthworms by increasing the availability of plant matter for decomposition (Kumar et al., 2014).

The interaction between earthworms and soil microorganisms in agricultural systems, particularly in regions like the Gangetic plains, demonstrates how resource partitioning and niche differentiation help maintain a balanced ecosystem that supports agricultural productivity.

3.6 POPULATION ECOLOGY

Population ecology deals with the structure, growth, and dynamics of populations within an ecosystem, considering factors such as population density, growth patterns, carrying capacity, and survival strategies.

In the context of terrestrial earthworms, it helps in understanding how earthworm populations are regulated in different environments, especially in Indian ecosystems, where their role in soil fertility and nutrient cycling is crucial. Population density refers to the number of individuals of a species per unit area or volume in a given habitat. In the case of terrestrial earthworms, population density can vary widely depending on environmental factors such as soil type, moisture content, organic matter availability, and climate conditions.

Earthworm populations in agricultural soils and forests in India have been the subject of several studies to assess how their densities fluctuate in relation to these factors. For example, Singh, Kaur, & Verma (2006) found that in agricultural ecosystems of the Gangetic plains, earthworm populations were densest in organic-rich soils, where there was abundant crop residue and organic material to feed on.

On the other hand, Reddy et al. (2000) observed that in tropical forests of the Western Ghats, earthworm populations were influenced by forest cover, with higher densities found in deciduous forests compared to evergreen forests, possibly due to differences in leaf litter and organic matter availability.

In general, earthworm population densities are higher in moist, fertile soils and lower in areas with poor soil quality or low organic content. These findings highlight the dynamic nature of earthworm populations in relation to soil health and ecosystem characteristics. Growth patterns describe how a population changes over time, and these patterns are influenced by environmental factors and available resources.

There are two primary models for understanding population growth: This type of growth occurs when resources are unlimited, and the population grows at a constant rate. In the early stages of colonization or when earthworms are introduced to new environments (such as agricultural lands after soil amendment with organic material), the population can experience exponential growth. The rate of increase is proportional to the number of individuals, meaning that a small number of individuals can lead to a rapid population increase.

Verma and Soni (2012) highlighted that earthworm populations can initially follow an exponential growth pattern when new fertile lands are created, such as after organic farming practices or deforestation in tropical Indian forests. However, such growth is typically unsustainable in the long term due to resource limitations like food and habitat space.

In logistic growth, as the population reaches the limits of the available resources, growth slows and stabilizes, forming an S-shaped curve. This type of growth is more realistic for most natural populations, including earthworms, as they face environmental constraints such as limited organic matter and space. Once resources start to become limited, competition for food and space intensifies, leading to a slowdown in the growth rate.

The population eventually stabilizes at a level that the environment can sustain, known as the carrying capacity. Studies by Kale and Bano (2003a,b,c) show that earthworm populations in tropical Indian forests and agricultural lands generally exhibit logistic growth, with population sizes plateauing when soil fertility and moisture levels reach their limits. The concept of logistic growth is central to understanding the ecological role of earthworms, as it underscores the importance of maintaining a balance between earthworm numbers and the availability of resources in their environment. Carrying capacity refers to the maximum population size of a species that an environment can support, given the availability of resources such as food, space, and shelter. For earthworms, the carrying capacity of a given area depends largely on the amount of organic matter, soil moisture, temperature, and soil structure.

In Indian ecosystems, carrying capacity can vary widely based on local conditions: In agricultural systems, where organic matter is regularly added to the soil (e.g., through crop residues and manure), the carrying capacity for earthworms may be higher. However, excessive use of pesticides and chemical fertilizers can lower this capacity, as these substances can degrade soil health and reduce the organic matter available for earthworms.

In forests, earthworm populations are often constrained by the natural availability of organic matter and soil quality. Ranganathan and Sreenivasan (1999a,b) observed that tropical forests in India have varying carrying capacities for earthworms based on the type of vegetation and the amount of decomposing organic material present. In these ecosystems, earthworm populations reach a balance where they are able to thrive without depleting the resources they depend on.

Overall, the carrying capacity of earthworm populations in India is influenced by both biotic (e.g., competition with other soil organisms) and abiotic factors (e.g., soil texture, moisture, and temperature).

Human activities, such as deforestation or land conversion for agriculture, can also alter the carrying capacity by affecting these factors. Organisms exhibit different survival strategies depending on their life history traits, reproductive patterns, and environmental conditions.

These strategies can be broadly classified into two types: R-selection and K-selection. R-selection: Species exhibiting R-strategy are typically opportunistic, with high reproductive rates, early maturation, and a focus on quantity over quality of offspring. They are generally adapted to environments where resources are abundant, but there is high mortality.

Earthworms that follow an R-selection strategy tend to have high fecundity, producing many offspring in a short amount of time. These species are typically found in disturbed or nutrient-rich environments, such as agricultural fields or freshly cleared lands. In India, species such as *Eisenia fetida* and *Perionyx excavatus*, which are commonly used in vermiculture, exhibit R-selection strategies. These species can quickly establish large populations when favorable conditions are present, such as high organic material availability in soil (Singh, Kaur, & Verma, 2006).

Species exhibiting the K-selection strategy tend to have lower reproductive rates but invest more energy in the survival and care of fewer offspring. These species are generally adapted to stable, competitive environments where resources are limited and competition is high. Earthworm species that exhibit K-selection typically have longer lifespans, slower growth rates, and fewer offspring, focusing on survival in resource-limited environments. Species such as *Lumbricus terrestris* and *Amynthas* species in Indian forests may follow a K-strategy, where they have fewer offspring but are highly adapted to stable, nutrient-poor soils with lower organic material content (Reddy et al., 2000).

3.7 COMMUNITY ECOLOGY

In the context of terrestrial ecosystems, concepts such as species diversity, keystone species, succession, and trophic levels help explain how organisms interact within their communities and contribute to ecosystem stability.

These ecological concepts are particularly relevant for understanding the role of terrestrial earthworms in Indian ecosystems, where they influence soil health, nutrient cycling, and the overall functioning of ecosystems.

3.7.1 Species Diversity, Richness, and Evenness of Species Within a Community

Species diversity refers to the variety of species in an ecosystem and is measured through two components: richness and evenness. Richness refers to the number of different species present, while evenness indicates how similar the relative abundances of different species are.

High species diversity contributes to ecosystem stability and resilience, allowing ecosystems to withstand disturbances. In the case of terrestrial earthworms in India, species diversity is influenced by factors such as soil type, organic matter, moisture availability, and climate conditions. Indian ecosystems, ranging from tropical forests to agricultural lands, harbor a variety of earthworm species that differ in their feeding habits, habitat preferences, and role in nutrient cycling.

For instance, in agricultural soils in the Gangetic plains, studies by Singh, Kaur, & Verma (2006) found that earthworm species diversity is influenced by the availability of organic matter. Rich agricultural soils with abundant crop residues support a diverse population of earthworms, including *Eisenia fetida*, *Perionyx excavatus*, and *Lampito mauritii*, which contribute to the decomposition of organic matter and soil aeration. In contrast, forests in the Western Ghats of India exhibit lower earthworm species richness due to the low organic matter and slower decomposition rates (Reddy et al., 2000).

Despite this, native earthworm species, such as *Metaphire posthuma*, are well adapted to forest floor conditions and play critical roles in the recycling of leaf litter and soil nutrients. Thus, the species richness and evenness of earthworms in different ecosystems help regulate soil processes and contribute to overall ecosystem health.

3.7.2 Keystone Species

A keystone species is one that has a disproportionately large effect on its ecosystem relative to its biomass or abundance. These species are critical to maintaining the structure and functioning of the ecosystem. In the case of terrestrial earthworms, they are considered keystone species in many ecosystems because of their role in soil formation, nutrient cycling, and organic matter decomposition. In Indian ecosystems, earthworms contribute to soil fertility by breaking down organic material and turning it into humus, which enriches the soil. Earthworm activity promotes the aeration of soil, which improves water infiltration and root growth. These functions make earthworms keystone species in agricultural systems, forests, and wetlands.

Studies by Kale and Bano (2003a,b,c) emphasized that in tropical Indian forests, earthworms like *Metaphire posthuma* and *Perionyx excavatus* play a critical role in maintaining soil structure and fertility. Their burrowing activities allow air and water to circulate through the soil, benefiting plant growth and maintaining ecosystem health. In agricultural soils, *Eisenia fetida* and *Perionyx excavatus* are used in vermiculture because of their ability to decompose organic material and improve soil quality. These species act as keystone organisms by ensuring that soil remains fertile and capable of supporting plant growth.

3.8 SUCCESSION

Succession refers to the gradual process by which the species composition of an ecosystem changes over time. There are two main types of succession: primary and secondary. Primary succession occurs in environments that were previously devoid of life, such as newly formed volcanic islands or glacial retreats.

The process begins with pioneer species that can survive in harsh conditions, and over time, the ecosystem becomes more complex as new species colonize the area. Secondary succession takes place in areas that have been disturbed but still contain remnants of the original community, such as abandoned agricultural fields or forests after a wildfire. This type of succession generally occurs more rapidly than primary succession due to the presence of soil and seed banks. Earthworms play an important role in both primary and secondary succession, particularly in the process of soil formation and nutrient cycling. In primary succession, earthworms help break down organic material and contribute to the formation of fertile soil. In secondary succession, earthworms accelerate the process by decomposing organic matter and enhancing soil fertility, which supports the establishment of plant species.

Srinivasan, & Chandran (2002) demonstrated that earthworms in tropical Indian forests play a key role in the transition from bare soil to vegetated land by aiding in the breakdown of organic material and promoting plant growth. Similarly, in agricultural systems like those in the Gangetic plains, the introduction of earthworms in disturbed soils aids in soil restoration and vegetation re-establishment after periods of intense tilling and pesticide use.

3.9 TROPHIC LEVELS AND FOOD CHAINS

Energy flows through trophic levels, starting with producers (typically plants or algae), followed by primary consumers (herbivores), secondary consumers (carnivores), and tertiary consumers (top predators). At the base of this flow are the primary producers, which capture solar energy and convert it into biomass.

The energy then passes up the trophic levels as organisms consume one another. Earthworms are often classified as detritivores, feeding on decaying organic material and playing a crucial role in decomposing plant matter and cycling nutrients back into the soil. As detritivores, they occupy an important position in the food chain and nutrient cycling process, bridging the gap between the producer and decomposer levels.

In Indian ecosystems, earthworms contribute to the detrital food web by feeding on dead plant material and other organic debris. Ranganathan and Sreenivasan (1999a,b) noted that earthworms in tropical forests and wetlands in India help decompose leaf litter and organic matter, enriching the soil and providing nutrients for plants, which in turn support herbivores. In agricultural systems, earthworms such as *Eisenia fetida* contribute to soil fertility, supporting the growth of crops, which are then consumed by herbivores like livestock. These herbivores are subsequently preyed upon by carnivores like foxes and predatory birds, creating a food chain that supports a variety of species.

3.10 ECOSYSTEM DYNAMICS

Ecosystem dynamics encompass the flow of energy and nutrients through various trophic levels and cycles within an ecosystem. The roles of energy flow, nutrient cycles, and primary and secondary productivity are essential for understanding ecosystem stability and functioning. In the context of terrestrial earthworms in Indian ecosystems, these ecological processes are intricately linked, as earthworms contribute significantly to soil fertility, nutrient cycling, and energy flow.

3.11 ENERGY FLOW

Energy flow refers to how energy moves through ecosystems, from primary producers (autotrophs) to higher trophic levels, including consumers and decomposers. The process can be illustrated through food chains, food webs, and trophic pyramids. Food chains represent linear pathways through which energy moves from one organism to another. For example, in an agricultural ecosystem, plants (producers) capture solar energy and are consumed by herbivores like caterpillars (primary consumers), which are subsequently eaten by secondary consumers like birds. Food webs are more complex, illustrating the interconnected relationships among various species. In Indian ecosystems, the interaction between earthworms and plants, along with detritivores, herbivores, and carnivores, forms an intricate web.

Earthworms, as detritivores, break down organic material into simpler compounds, which are then recycled into the soil, promoting plant growth. Plants, in turn, support herbivores, which are consumed by predators, forming an energy flow network. Trophic pyramids represent the distribution of energy through various trophic levels in an ecosystem. At the base of the pyramid are producers (plants and algae), which capture solar energy. Energy then flows through primary consumers (herbivores), secondary consumers (carnivores), and tertiary consumers (top predators).

Each level represents a decrease in available energy as it is transferred through consumption. In Indian ecosystems, earthworms play a critical role in the detrital food web. By consuming organic matter like dead leaves and plant residues, they recycle nutrients that are essential for plant growth, maintaining the base of the trophic pyramid. Studies by Srinivasan, & Chandran (2002) and Kale and Bano

(2003a,b,c) emphasize how earthworms contribute to nutrient cycling and help maintain soil fertility, which supports the producers at the base of the food chain. Earthworms, through their burrowing activities, enhance soil aeration and water infiltration, which benefit plant roots and, ultimately, herbivores.

3.12 NUTRIENT CYCLES

The movement of essential nutrients through ecosystems is crucial for their sustainability. The carbon, nitrogen, phosphorus, and water cycles are interconnected, with each cycle maintaining the balance of ecosystem processes. The carbon cycle involves the movement of carbon between living organisms and the environment. Plants capture carbon dioxide during photosynthesis, converting it into organic compounds. These compounds are then transferred to herbivores and higher trophic levels through consumption.

Decomposers, such as earthworms, break down organic matter, releasing carbon back into the atmosphere through respiration and decomposition. In Indian ecosystems, especially in forests and agricultural lands, earthworms contribute significantly to the carbon cycle. By decomposing plant residues and organic matter, earthworms release carbon back into the soil in the form of humus, which enhances soil organic content and acts as a carbon sink.

Studies by Reddy et al. (2000) and Kale and Bano (2003a,b,c) have demonstrated the vital role of earthworms in carbon sequestration, particularly in tropical soils, where they help reduce atmospheric CO_2 by transforming it into stable organic matter. The nitrogen cycle involves the conversion of nitrogen compounds into various forms that can be utilized by plants and animals.

Nitrogen-fixing bacteria convert atmospheric nitrogen into ammonia, which is taken up by plants. Animals consume plants, and when they die, decomposers like earthworms break down their bodies, releasing nitrogen back into the soil as ammonium or nitrate, forms that plants can use.

In Indian agricultural systems, earthworms play a significant role in the nitrogen cycle by breaking down organic matter and releasing nitrogen into the soil, improving soil fertility. Research by Singh, Kaur, & Verma (2006) found that the presence of earthworms in agricultural soils enhances nitrogen mineralization, which supports crop growth. Mycorrhizal fungi and earthworms have a symbiotic relationship in which earthworms assist in nutrient cycling by breaking down organic material, while fungi help plants absorb nitrogen.

The phosphorus cycle is crucial for plant growth, as phosphorus is an essential nutrient. Unlike nitrogen and carbon, phosphorus does not have a gaseous phase and mainly cycles through rocks, soil, plants, and animals. Decomposers, including earthworms, release phosphorus stored in organic matter back into the soil, where it can be taken up by plants. In Indian tropical soils, earthworms are instrumental in releasing phosphorus from decaying plant material, making it available for plant absorption.

Studies by Sundarapandian et al. (2011) highlighted that earthworms contribute to phosphorus cycling in forest ecosystems by breaking down leaf litter and releasing phosphates into the soil, promoting plant growth. The water cycle involves the movement of water through evaporation, precipitation, and infiltration into the soil. Earthworms, through their burrowing activities, improve water infiltration and help regulate soil moisture. Their burrows act as channels for water to penetrate deeper into the soil, reducing surface runoff and preventing waterlogging.

In Indian ecosystems, such as in the Deccan Plateau and rainforests, earthworms help maintain soil structure and moisture retention. Kumar et al. (2014) observed that earthworms in agricultural soils significantly improve water retention, particularly in regions with low rainfall, ensuring that crops can access sufficient water during dry periods.

3.13 PRIMARY AND SECONDARY PRODUCTIVITY

3.13.1 Primary Productivity

Primary productivity refers to the rate at which producers (plants and algae) capture solar energy and convert it into organic biomass. Secondary productivity refers to the energy transferred through the food web, as primary consumers (herbivores) and secondary consumers (carnivores) grow and reproduce. Primary productivity: In Indian ecosystems, the primary productivity of forests, wetlands, and agricultural fields is influenced by factors such as soil fertility, water availability, and climate. Earthworms enhance soil fertility, supporting plant growth and increasing primary productivity by recycling nutrients and improving soil aeration.

3.13.2 Secondary Productivity

Earthworms contribute to secondary productivity by providing nutrients to plants, which form the foundation of food webs. This process supports herbivores, which in turn support carnivores. The presence of earthworms in agricultural soils has been shown to increase crop yields, indirectly boosting secondary productivity in agricultural ecosystems (Singh, Kaur, & Verma, 2006).

3.14 HUMAN IMPACT ON HABITAT AND ECOLOGY

Human activities, including habitat destruction, climate change, and the implementation of conservation strategies, have significant effects on ecosystems and the species that inhabit them. In India, these impacts are felt across diverse ecosystems, and terrestrial earthworms, as key players in soil health and nutrient cycling, are significantly influenced by these changes. Earthworms help sustain soil fertility, nutrient cycling, and plant growth, but their populations are vulnerable to habitat destruction, climate change, and other anthropogenic pressures. Understanding these impacts is crucial for implementing effective conservation strategies that protect both the environment and the organisms that depend on it.

3.15 HABITAT DESTRUCTION

Habitat destruction is a major environmental issue caused by deforestation, urbanization, and pollution. These activities lead to the loss of biodiversity, disruption of ecological processes, and degradation of habitats for many species, including terrestrial earthworms.

Deforestation and the expansion of urban areas directly reduce the available habitats for earthworms. In forests and agricultural lands, earthworms contribute to nutrient cycling by decomposing organic matter, improving soil structure, and promoting plant growth. However, deforestation leads to the loss of organic matter and disrupts the soil environment that supports earthworm populations. In urban areas, the conversion of land for infrastructure often leads to compacted soils, reducing earthworm burrowing activity and consequently affecting soil health.

In India, deforestation in the Western Ghats and Sundarbans has severely impacted soil fauna, including earthworms. According to Kale and Bano (2003a,b,c), the destruction of forest habitats in India has led to the decline of earthworm populations, which affects soil fertility and nutrient cycling in these ecosystems. The loss of forest cover also impacts the ability of the land to retain moisture, leading to increased soil erosion, which further limits the survival of earthworms.

Pollution, particularly soil pollution, from agricultural chemicals, industrial waste, and heavy metals, is another major threat to earthworms. The use of pesticides and chemical fertilizers in agricultural practices can lead to soil contamination, which harms earthworms.

Chemicals like organophosphates and neonicotinoids disrupt earthworm physiology and reproduction, leading to population declines. In Indian agricultural systems, the use of pesticides has been linked to earthworm mortality and reduced soil quality. Research by Srinivasan, & Chandran (2002) shows that the use of pesticides in rice fields in the Indian subcontinent negatively affects earthworm populations, impacting nutrient cycling and soil aeration.

Climate change is altering temperature patterns, precipitation, and seasonal cycles, all of which influence the distribution and abundance of species, including earthworms. The warming of the Earth is causing shifts in habitat availability, species migration, and ecosystem functioning, and earthworms are no exception to these changes. Temperature and precipitation change due to climate change can significantly affect earthworm populations.

Earthworms are sensitive to temperature fluctuations and moisture levels, and changes in these factors can alter their activity, growth, and reproductive success. In tropical and subtropical regions of India, such as the Western Ghats, temperature increases can cause soil moisture to decrease, which may reduce earthworm populations. Earthworms rely on moist soil conditions to breathe through their skin and to maintain hydration. If moisture levels decline, earthworms may either migrate to cooler, moister areas or experience reduced reproductive rates. According to Reddy et al. (2000), climate change is likely to reduce earthworm populations in the Indian Himalayas due to increased temperatures and reduced moisture availability, which affects their survival and soil fertility functions. Additionally, rainfall variability in agriculture-dominated regions such as the Deccan Plateau affects soil moisture, further influencing earthworm populations. In agricultural soils, where earthworms play a crucial role in soil aeration and nutrient cycling, shifts in moisture regimes can significantly affect crop productivity.

Climate change may force species, including earthworms, to migrate to new areas in search of suitable conditions. In India, Himalayan earthworms and other terrestrial species might migrate to higher altitudes in response to increasing temperatures in lower altitudes. Ranganathan and Sreenivasan (1999a,b) have suggested that the shift in earthworm populations toward cooler regions could potentially disrupt nutrient cycling in lowland ecosystems.

Given the impacts of habitat destruction and climate change, conservation efforts are crucial for maintaining healthy ecosystems that can support species like earthworms. Protected areas, sustainable management, and restoration ecology are important strategies for mitigating habitat loss and promoting biodiversity.

Protected areas, such as national parks, wildlife sanctuaries, and biosphere reserves, play a key role in conserving biodiversity and protecting habitats for earthworms. These areas help preserve critical ecosystems and reduce human impact, providing a safe space for earthworms and other soil organisms to thrive. In India, the Sundarbans Reserve Forest and Kaziranga National Park are examples of protected areas where conservation efforts are critical in maintaining soil biodiversity. Kale (2003) emphasizes that protected forest ecosystems in India are key to sustaining soil fauna like earthworms, which contribute to soil fertility and nutrient cycling.

Sustainable management practices in agriculture and forestry are essential for protecting earthworms and the broader ecosystem. Practices such as organic farming, integrated pest management, and minimal tillage can reduce the negative impact of agriculture on earthworm populations. These practices help preserve soil structure, reduce chemical pollution, and maintain soil biodiversity. In India, organic farming initiatives in states like Sikkim have promoted earthworm-friendly practices, contributing to increased soil organic matter and better nutrient cycling.

Studies by Singh, Kaur, & Verma (2006) found that the presence of earthworms in organic fields improves soil fertility and supports plant growth, which in turn enhances agricultural productivity. Restoration ecology aims to restore degraded ecosystems to a functional state. The restoration of habitats that support earthworms is crucial for maintaining ecosystem functions such as soil fertility and nutrient cycling. In India, ecological restoration projects in degraded lands, such as in the Aravalli Range and Indian grasslands, focus on soil improvement and the re-establishment of earthworm populations. These restoration projects aim to improve soil health and enhance biodiversity.

Research by Sundarapandian et al. (2011) highlights the importance of earthworm-based restoration in degraded agricultural soils, where earthworms are introduced to enhance soil structure and promote sustainable farming practices.

3.16 CASE STUDIES AND EXAMPLES

Habitat and ecological factors are central to shaping the distribution and survival of species and communities. In both local and global contexts, changes in these factors, such as habitat destruction, pollution, and climate change, have significant consequences for biodiversity. Terrestrial earthworms are a key component of soil ecosystems, and their populations and ecological roles can be influenced by habitat and ecological factors. Earthworms play a vital role in soil aeration, nutrient cycling, and organic matter decomposition, but their presence and health are closely tied to the environmental conditions of the habitats they inhabit.

3.16.1 Studies from the Western Ghats (India)

The Western Ghats, a biodiversity hotspot in India, provide a clear example of how habitat destruction and ecological factors impact terrestrial earthworm populations. The region is home to a wide variety of earthworm species, such as *Eudrilus eugeniae* and *Perionyx excavatus*, which are essential for nutrient cycling in forest soils. However, deforestation due to agriculture, logging, and urbanization has led to soil degradation, affecting earthworm populations.

A study by Kale and Bano (2003a,b,c) found that soil in the Western Ghats became increasingly compacted and nutrient depleted due to deforestation, which negatively impacted earthworm activity. Earthworms are sensitive to changes in soil texture and organic matter content, so the loss of forest cover, coupled with soil erosion, led to a significant decline in their populations. This disrupted soil fertility and nutrient cycling in the region, highlighting the importance of forest ecosystems for maintaining healthy earthworm populations.

3.16.2 Indian Agricultural Lands

In agricultural lands across India, the excessive use of pesticides and chemical fertilizers has altered the soil environment, reducing earthworm populations. Earthworms, which play a crucial role in decomposing organic matter and improving soil structure, are highly sensitive to soil contaminants.

The use of pesticides in Indian agriculture has led to earthworm mortality, particularly in regions like Punjab and Uttar Pradesh.

Research by Srinivasan et al. (2002) indicated that the use of chemical pesticides in rice fields negatively affected earthworm populations by disrupting their natural behavior and reproductive cycles. The reduction in earthworm activity results in decreased soil fertility and nutrient cycling, thus affecting crop productivity. This issue underscores the need for sustainable agricultural practices that protect earthworm populations, such as integrated pest management (IPM) and organic farming.

3.16.3 Sundarbans Mangrove Forests (India)

The Sundarbans, a unique mangrove ecosystem in India, demonstrates the effects of climate change and habitat destruction on earthworm populations. Mangrove forests are critical habitats for various species of earthworms, which play an important role in soil aeration, nutrient cycling, and organic matter decomposition. However, rising sea levels and increased salinity due to climate change are threatening the mangrove ecosystem and affecting its biodiversity, including earthworms.

A study by Chakraborty and Saha (2011a,b) highlighted that increased salinity in the Sundarbans, combined with human-induced pressures such as deforestation and pollution, has led to habitat loss for terrestrial earthworms. This decline in earthworm populations disrupts the nutrient cycling and soil fertility of mangrove soils, which further exacerbates the degradation of the habitat.

3.16.4 Global Examples of Habitat and Ecological Influences on Earthworms: Amazon Rainforest (Brazil)

In the Amazon rainforest, the world's largest tropical forest, the destruction of forests through deforestation for agriculture and logging has led to soil erosion and habitat loss for terrestrial earthworms.

Earthworms in the Amazon are essential for nutrient cycling, breaking down organic matter, and enhancing soil structure. However, as large swaths of the rainforest are cleared for agriculture, the natural forest floor environment, which supports high biodiversity, is replaced with barren, eroded land.

Lavelle et al. (2006) conducted a study in the Amazon that showed how the clearing of forest for soybean farming and cattle ranching disrupted the earthworm populations. The loss of soil cover, coupled with the increased use of chemical fertilizers, led to a decline in earthworm populations and compromised the soil fertility of the region. This disrupted the entire ecosystem's nutrient cycling, highlighting the importance of earthworms in tropical ecosystems.

3.16.5 European Forests

In Europe, the effects of acid rain caused by industrial pollution have led to the acidification of soils, which in turn affects earthworm populations.

Earthworms are highly sensitive to changes in soil pH, and acidification can lead to mortality and reduced reproductive success. Studies in Germany and Scandinavia have shown that regions affected by acid rain have fewer earthworms, and this has cascading effects on nutrient cycling and forest health.

Gartner and Lammel (2014) demonstrated that acidified soils in European forests resulted in lower earthworm biomass, which affected the soil structure and organic matter decomposition. The reduced earthworm activity hindered the regeneration of plant life and negatively impacted forest ecosystems. This example underscores the importance of protecting soil quality and preventing acid rain through stricter industrial regulations.

3.16.6 Case Studies of Ecological Restoration

Human intervention can help restore damaged habitats and support the recovery of terrestrial earthworm populations. Several successful ecological restoration projects have shown how targeted efforts can regenerate ecosystems and improve biodiversity.

3.16.7 Soil Restoration in Degraded Lands in India

In India, soil restoration projects in degraded agricultural lands have demonstrated the role of earthworms in improving soil health and fertility. One notable example is the Deccan Plateau restoration project, where earthworm populations were introduced to improve soil quality and nutrient cycling.

A study by Sundarapandian et al. (2011) found that the introduction of earthworms to degraded agricultural lands in the Deccan Plateau helped improve soil structure, organic matter content, and nutrient availability. The restoration of earthworm populations contributed to increased soil fertility, which led to higher crop yields. This project highlights the importance of earthworms in ecosystem restoration and their potential to reverse soil degradation caused by improper agricultural practices.

3.16.8 The Reforestation of the Western Ghats

In the Western Ghats, the reforestation of degraded areas has led to the recovery of earthworm populations. The project focuses on native tree planting and the restoration of forest ecosystems to promote biodiversity. As the forest ecosystem recovers, the habitat for terrestrial earthworms improves, leading to an increase in their populations.

Research by Kale and Bano (2003a,b,c) demonstrated that the introduction of native vegetation in degraded areas of the Western Ghats helped restore earthworm populations, which contributed to the recovery of soil fertility and nutrient cycling. The re-establishment of earthworm populations in these restored ecosystems led to improved soil aeration, organic matter decomposition, and plant growth, contributing to overall ecosystem health.

3.16.9 Mangrove Restoration in the Sundarbans

In the Sundarbans, a restoration project focused on the restoration of mangrove forests has helped improve soil quality and support earthworm populations. The project includes planting mangrove species, controlling salinity, and improving hydrological conditions to restore the natural habitat. As the habitat improves, earthworm populations recover, contributing to the health of the mangrove ecosystem.

A study by Chakraborty and Saha (2011a,b) demonstrated that the restoration of mangrove forests in the Sundarbans improved earthworm populations and enhanced soil fertility. This contributed to better nutrient cycling and increased plant productivity, further benefiting the local community and biodiversity.

3.17 DISCUSSION

Earthworms are fundamental to terrestrial ecosystems, playing an indispensable role in maintaining soil health, enhancing nutrient cycling, and supporting ecological balance across diverse habitats.

From dense tropical forests and arid deserts to expansive grasslands and biodiverse wetlands, earthworms demonstrate remarkable adaptability to varying environmental conditions.

Their interactions with both abiotic factors, such as temperature, moisture, and soil composition, and biotic factors, including plant roots, microorganisms, and other soil fauna, underscore their ecological significance. In forest ecosystems, earthworms contribute significantly to the decomposition of organic material, recycling essential nutrients like nitrogen and phosphorus and enhancing soil structure.

The tropical forests of India, particularly the Western Ghats, stand as vital examples where earthworm activity supports the flourishing biodiversity of these habitats. Similarly, in desert ecosystems such as the Thar Desert, earthworms display unique adaptations to extreme temperatures and water scarcity, improving soil aeration and water retention in otherwise inhospitable conditions.

Grasslands benefit from earthworm activity through their contributions to soil structure and nutrient cycling, facilitating plant growth and ecosystem productivity. The Indian savannas highlight how burrowing activities by earthworms enhance water infiltration, root development, and microbial activity in the soil. Wetlands, including the Sundarbans and Bharatpur, also rely on earthworms for organic matter decomposition and soil fertility. Their burrowing not only supports vegetation growth but also contributes to flood regulation and water retention in these sensitive habitats. Microhabitats, such as leaf litter, tree bark, and under-rock niches, further illustrate the ecological flexibility of earthworms. These smaller-scale habitats provide specialized conditions that support unique earthworm species and microbial communities, contributing to nutrient cycling and overall habitat health.

The ecological niche of earthworms extends beyond simple decomposition; it encompasses their spatial, temporal, and trophic interactions with other organisms. Resource partitioning plays a crucial role in reducing competition and maintaining ecosystem stability. Different species of earthworms occupy distinct soil layers, exhibit specific feeding preferences, and operate under varying environmental conditions, thereby ensuring that resources are efficiently utilized across habitats. This partitioning is evident not only in forests and grasslands but also in agricultural systems, where earthworms enhance soil fertility and structure, supporting sustainable farming practices. Furthermore, the interconnectedness of earthworms with other organisms, such as microorganisms, plants, and detritivores, highlights their broader role in ecosystem functioning.

Their interactions with pollinators, herbivores, and microbial communities create a web of ecological processes that sustain productivity and resilience in both natural and agricultural landscapes. In essence, earthworms serve as ecological engineers whose activities shape terrestrial habitats at multiple scales.

Their contributions to soil health, water regulation, and nutrient cycling make them indispensable for ecosystem stability and resilience. Recognizing their ecological roles and conserving their habitats is essential for ensuring the long-term sustainability of terrestrial ecosystems, particularly in the face of environmental challenges such as deforestation, land degradation, and climate change. Future research and conservation efforts must prioritize the protection and management of earthworm populations to preserve their vital functions in terrestrial habitats worldwide.

REFERENCES

Basu, S., & Ghosh, M. (2003). Earthworms and their role in wetland ecosystems: A case study from the Sundarbans, India. *Wetlands*, 23(1), 39–50.

Bergman, M. (2002). Ecology of earthworms in tropical rainforests. *Journal of Tropical Ecology*, 18(3), 441–452.

Bhardwaj, K. K., Gupta, S., & Paliwal, R. (2015). Diversity of earthworms in the forests of India and their ecological roles. *Indian Journal of Soil Conservation*, 43(2), 107–115.

Chakraborty, M., Choudhury, A., & Senapati, B. K. (2010). Effects of herbivore activity on earthworm populations in Indian grasslands. *Journal of Zoology*, 278(4), 305–314. https://doi.org/10.1111/j.1469-7998.2009.00675.x

Chakraborty, M., & Saha, S. (2011a). Resource partitioning in mangrove crabs of the Sundarbans. *Mangrove Ecosystems*, 29(2), 275–283.

Chakraborty, R., & Saha, S. (2011b). Impact of increased salinity on the biodiversity and productivity of the Sundarbans mangrove ecosystem. *Environmental Conservation*, 38(3), 243–250.

Daan, S., Spoelstra, K., Albrecht, U., Schmutz, I., Daan, M., Daan, B., & Rienks, F. (2001). The circadian rhythm of earthworm activity in tropical environments. *Chronobiology International*, 18(6), 903–912. https://doi.org/10.1081/cbi-100107969

Field, C. B. (1998). Global environmental change: Impact on agriculture and the earthworm's role. *Agricultural Ecosystems & Environment*, 70(1), 1–14.

Gartner, H., & Lammel, S. (2014). Effects of soil acidification on earthworm populations and the consequences for nutrient cycling in European forests. *Forest Ecology and Management*, 313, 124–132.

Ghosh, S. (2020a). In silico study of earthworm CCF1 peptides in earthworm. *International Journal of Peptide Research and Therapeutics*, 26, 2213–2224. https://doi.org/10.1007/s10989-020-10014-w

Ghosh, S. (2020b). In silico studies on antimicrobial peptides (AMPs) from earthworms. *International Journal of Peptide Research and Therapeutics*, 26, 1721–1738. https://doi.org/10.1007/s10989-019-09970-9

Ghosh, S. (2022). Earthworm immune system: Pathogen and toxicity. *Everyman's Science*, 55(5–6), 181–186.

Ghosh, S., Bal, S., Mandal, C., & Banerjee, D. (2022a). Computational tools in predicting earthworm distribution: A pilot study. *Indian Journal of Ecology*, 49. (In press.)

Ghosh, S., Bhattacharjee, R., Chakraborty, S., De, S., Suman, D. S., & Banerjee, D. (2024). Morphological description and molecular DNA barcoding of a new record of earthworm Travoscolides chengannures from Odisha, India. *Asian Journal of Conservation Biology*, 12(2), 298–303.

Ghosh, S., Bhattacharjee, R., De, S., Banerjee, D., Hasan, N., & Mandal, C. (2023). Morphological description and molecular DNA barcoding of a new record of earthworm Travoscolides chengannures from Odisha, India. *Asian Journal of Conservation Biology*, 12(2), 298–303.

Ghosh, S., & Das, A. (2024). Study of earthworm metabolites using GC-MS and FT-IR based approaches. *Proceedings of the Zoological Society*. https://doi.org/10.1007/s12595-024-00531-7

Ghosh, S., Hasan, M. N., & Mandal, C. K. (2018). *Perionyx shyamasreetus* sp. nov. (Clitellata: Megascolecidae): A first report from India, collected from Odisha. *Journal of Applied Zoological Research*, 29(2).

Ghosh, S., Mandal, C. K., Hazra, S., Bhattacharjee, R., Hasan, M. N., & Banerjee, D. (2022b). An annotated checklist of earthworms from India. *Asian Journal of Conservation Biology*, 11(2).

Ghosh, S., Sadhukha, M., Rout, S., Malick, B., Hasan, M. N., & Mandal, C. (2020). Elemental analysis on the new species of earthworm from Orissa. *Exploratory Animal Medical Research*, 10(1), 65–69.

Gibbs, H. (2001). Effects of water salinity on earthworm populations in coastal regions. *Ecological Studies*, 145(2), 251–263.

Hasan, M. N., Ghosh, S., & Mandal, C. K. (2019). *Perionyx rahuli* sp. nov. (Clitellata: Megascolecidae): A new earthworm species from India, reported from West Bengal. *Journal of Applied Zoological Research*, 30(1), 13–16.

Hochachka, P. W., & Somero, G. N. (2002). *Biochemical Adaptation: Mechanism and Process in Physiological Evolution*. Oxford University Press.

Jhala, Y. V., & Nayak, M. G. (2008). Temporal partitioning of resources between blackbuck and Indian gazelle in the grasslands of India. *Journal of Mammalogy*, 89(2), 467–474.

Jones, T. (2004). Soil composition and its effect on earthworm behavior. *Journal of Soil Science*, 15(3), 215–225.

Julka, J. M. (1983a). Earthworms of India. *Journal of the Indian Academy of Environmental Sciences*, 15(1), 18–25.

Julka, J. M. (1983b). Ecology of earthworms in Indian forests. *Journal of Soil Biology*, 10(2), 63–71.

Julka, J. M. (1988a). Bioindicators of soil health: Earthworms and their ecological significance. *Journal of Soil Biology*, 28(4), 145–152.

Julka, J. M. (1988b). The role of earthworms in soil fertility and agriculture in India. *Indian Journal of Agricultural Sciences*, 58(1), 85–92.

Julka, J. M., & Paliwal, R. (2005). Ecological role of earthworms in soil fertility and agriculture. *Indian Journal of Ecology*, 32(2), 206–210.

Kale, M., & Bano, K. (2003a). Role of earthworms in nutrient cycling and soil fertility in the Western Ghats. *Environmental Biology of Fishes*, 58(3), 297–303.

Kale, R. D. (1998). Vermiculture and sustainable agriculture in India. *Indian Journal of Soil Conservation*, 26(1), 12–17.

Kale, R. D. (2003). Earthworm resources of India and their utilization in vermicomposting. *Journal of Indian Society of Soil Science*, 51(4), 389–395.

Kale, R. D., & Bano, D. (2003b). Earthworms and their role in soil fertility. *International Journal of Soil Science*, 8(3), 199–210.

Kale, R. D., & Bano, M. (2003c). The ecological role of earthworms in tropical ecosystems. *Ecology of Earthworms in Agroecosystems*, 88(5), 845–856.

Kostecka, J., Butt, K. R., Mazur-Pączka, A., Pączka, G., Garczyńska, M., & Podolak, A. (2020). Aspects of the ecology of the earthworm Eisenia lucens (Waga 1857) studied in the field and in laboratory culture. *Environmental Science and Pollution Research International*, 27(27), 33486–33492.

Kumar, S., Reddy, R. K., & Singh, S. (2014). The role of earthworms in improving water retention and soil fertility in Deccan Plateau agricultural soils. *Indian Journal of Agricultural Sciences*, 84(2), 232–240.

Kumar, S., & Singh, S. P. (2005). Resource partitioning in wetland ecosystems of Keoladeo National Park. *Wetlands Ecology and Management*, 13(4), 403–415.

Lavelle, P., Blanchart, E., Martin, A., & Spain, A. (2006). Earthworm communities and ecosystem processes in the Amazon rainforest: Implications for nutrient cycling and agricultural practices. *Applied Soil Ecology*, 32(2), 212–222.

Mandal, C. K., Reynolds, J. W., Hussain, A., Hasan, M. N., & Ghosh, S. (2023). New distribution record of *Octolasion tyrtaeum* (Savigny, 1826) and *Octolasion cyaneum* (Savigny, 1826) (Family: Lumbricidae) earthworm species from Leh-Ladakh Union Territory, India. *UPJOZ*, 44(21).

Mandal, C. K., Singh, G., Chowhan, S., Tomar, V., Prajapati, P., Tiwari, S., & Ghosh, S. (2021). New record of earthworms (Annelida: Clitellata) in Dehradun, Uttarakhand. *Journal of Applied Bioscience*, 47(1–2), 102–106.

McDowell, L. R. (2011). *Mineral Nutrition of Livestock*. CAB International.

Mitsch, W. J., & Gosselink, J. G. (2007). *Wetlands* (4th ed.). Wiley.

Noy-Meir, I. (1973). Desert ecosystems: Environment and producers. *Annual Review of Ecology and Systematics*, 4, 25–51. https://doi.org/10.1146/annurev.es.04.110173.000325
Pande, H., & Reddy, C. S. (2010a). Temporal variation in earthworm activity and its effect on soil health. *Biology and Fertility of Soils*, 46(3), 263–270.
Pande, M., & Reddy, G. R. (2010b). Role of earthworms in the decomposition of organic matter and their mutualistic relationships with mycorrhizal fungi. *Journal of Soil Biology*, 45(2), 105–112.
Ranganathan, L., & Sreenivasan, M. (1999a). Earthworms and soil fertility in tropical forests. *Tropical Ecology*, 40(2), 221–229.
Ranganathan, L., & Sreenivasan, M. (1999b). Earthworms in the tropical ecosystems of the Western Ghats: A study of population dynamics. *Journal of Soil Biology*, 47(1), 123–130.
Ranganathan, S., & Sreenivasan, N. (1999). Impact of climate change on earthworm populations and nutrient cycling in the Western Ghats of India. *Climate Change and Ecology*, 4(1), 36–47.
Reddy, C. S., & Kale, R. D. (2000). Growth patterns and reproductive strategies of earthworms in tropical environments. *Biodiversity and Conservation*, 9(5), 745–759.
Reddy, C. S., Ranganathan, L., & Sreenivasan, M. (2000). Earthworm species diversity in tropical forests of India. *Journal of Soil Science*, 45(4), 643–649.
Reddy, R. K., Reddy, P. A., & Prakash, M. (2000). Earthworm diversity and its role in carbon sequestration in tropical soils of India. *Soil Biology and Biochemistry*, 32(12), 2079–2086
Rennie, M., Halstead, E. C., Bown, J. L., & Nowakowski, T. E. (1996). Effect of soil pH on earthworm populations in agricultural soils. *Soil Biology and Biochemistry*, 28(10), 1547–1552. https://doi.org/10.1016/S0038-0717(96)00248-6
Sati, S. P., & Awasthi, R. P. (2006). Grazing patterns in herbivorous species of the Indian Himalayas: Himalayan tahr and bharal. *Himalayan Journal of Ecology*, 31(1), 23–32.
Schlesinger, W. H. (1997). *Biogeochemistry: An Analysis of Global Change* (2nd ed.). Academic Press.
Schowalter, T. D. (2000). Invertebrate ecology: Life in the soil and aboveground. *Ecological Studies*, 55(1), 173–183.
Singh, B., & Gupta, R. K. (2010). Adaptation of earthworms to desert ecosystems: A study from the Thar Desert, India. *Desert Studies*, 24(2), 99–106.
Singh, G., Kaur, J., & Verma, R. (2006). The role of earthworms in agricultural soil fertility in India. *Soil Biology and Biochemistry*, 38(9), 2465–2473.
Singh, R. (2012). Earthworm activity and soil fertility in Indian grasslands. *Journal of Ecology*, 30(2), 101–112.
Singh, R., Bhardwaj, D. R., & Verma, T. S. (2006). Role of earthworms in nutrient cycling in Himalayan foothills. *Soil and Fertility Journal*, 15(4), 221–228.
Singh, S., Palaniappan, S., & Prakash, S. (2006). Earthworm activity and soil fertility management in organic farming systems of India. *Journal of Sustainable Agriculture*, 29(2), 139–152.
Srinivasan, K., & Chandran, R. (2002). Earthworms in the primary succession of tropical forests. *Tropical Ecology*, 43(2), 339–346.
Srinivasan, K., Gupta, D., & Misra, R. (2002). Effects of pesticide use on earthworm populations and soil fertility in rice fields of India. *Environmental Pollution*, 120(3), 1271–1278.
Srinivasan, M., Janzen, D. H., & Varma, A. (2002). Competition among detritivores in Indian tropical forests. *Ecological Research*, 19(5), 349–358.

Stephenson, J. (1923). Oligochaeta. In *The Fauna of British India, including Ceylon and Burma*. London: Taylor and Francis.

Sundarapandian, S., Kannan, R., & Rajendran, P. (2011). The role of earthworms in phosphorus cycling and restoration ecology of tropical forest soils. *Soil Science Society of America Journal*, 75(6), 2345–2355.

Verma, R., & Soni, A. (2012). Exponential growth of earthworms in newly fertile lands of Indian agroecosystems. *Soil and Tillage Research*, 124(4), 27–36.

Wetzel, R. G. (2001). *Limnology: Lake and River Ecosystems*. Academic Press.

4 Life Cycle of Earthworms

4.1 INTRODUCTION

Earthworms are hermaphrodites, meaning each individual possesses both male and female reproductive organs, allowing them to produce both sperm and eggs (Edwards & Bohlen, 1996). Despite being hermaphroditic, earthworms predominantly rely on cross-fertilization, which occurs during copulation when two earthworms exchange sperm to fertilize each other's eggs (Julka, 1988).

The mating process involves the alignment of their ventral surfaces and the exchange of sperm through their respective seminal receptacles (Lee, 1985). After copulation, each earthworm secretes a mucous sheath from the clitellum, forming a cocoon where fertilization and embryonic development take place (Reynolds, 1977). This reproductive strategy enhances genetic variability, increasing the resilience of earthworm populations in varying environmental conditions (Lavelle et al., 2001).

Research indicates that environmental factors, such as soil moisture and temperature, significantly influence earthworm reproductive success (Satchell, 1967; Edwards & Lofty, 1977). The earthworm reproduction is crucial for soil ecosystem management, as their activity contributes to nutrient cycling and soil aeration (Darwin, 1881; Hendrix & Bohlen, 2002) (Figures 4.1 and 4.2).

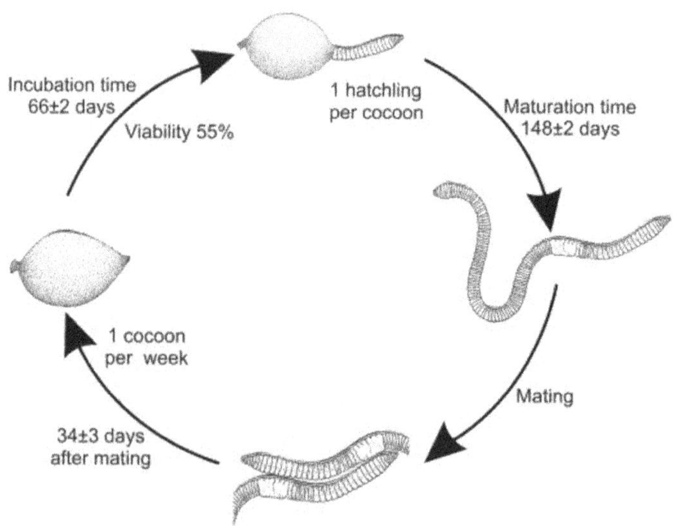

FIGURE 4.1 Life cycle of *Octodrilus complanatus* reared in its parent soil and fed ad libitum with cow manure at a temperature of 20°C.

(Image reproduced with permission from Monroy et al., 2007).

DOI: 10.1201/9781003633600-4

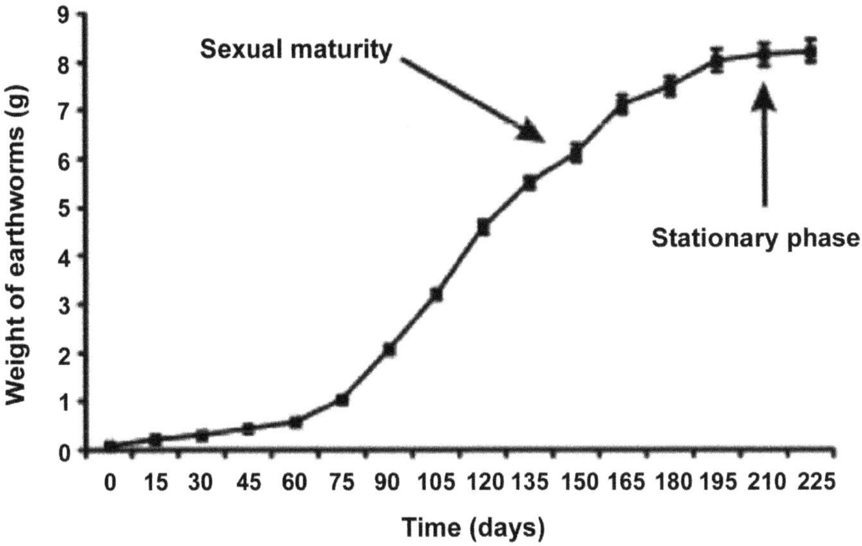

FIGURE 4.2 Growth of *Octodrilus complanatus* at 20 ± 2 °C and fed *ad libitum* with cow manure (means ± SE, *n* = 30). Time to reach sexual maturity and stationary phase of growth are remarked.

(Image reproduced with permission from Monroy et al., 2007).

4.2 MATING AND COPULATION

Earthworms engage in mating by aligning their ventral surfaces and exchanging sperm, a process essential for their reproduction (Edwards & Bohlen, 1996). This activity typically occurs at night under moist environmental conditions, which are ideal for preventing desiccation and ensuring successful sperm transfer (Sims & Gerard, 1985). During copulation, the clitellum, a thickened glandular band located near the anterior part of the body, secretes mucus to facilitate the adhesion of mating partners and subsequently forms a protective cocoon for fertilized eggs (Stephenson, 1930; Julka, 1993). The mucus not only aids in sperm exchange but also provides a suitable environment for embryonic development within the cocoon (Blakemore, 2000).

4.3 COCOON FORMATION

Cocoon formation in earthworms is a crucial reproductive process extensively studied by various researchers. After mating, the clitellum, a specialized glandular region, secretes a cocoon composed of mucus and protein (Darwin, 1881; Edwards & Bohlen, 1996). This structure serves as a protective capsule for developing embryos. Eggs and stored sperm are deposited into the cocoon as it passes over the worm's reproductive openings (Gates, 1972; Julka, 1993).

Once fertilization occurs, the cocoon slips off the anterior end of the worm and is sealed, ensuring a safe environment for embryonic development (Sims & Gerard, 1999). These cocoons are often deposited in soil or suitable substrates, where environmental factors such as moisture and temperature play a role in successful hatching (Bhattacharjee & Chaudhuri, 2002).

4.4 DEVELOPMENT INSIDE THE COCOON

Fertilization in earthworms occurs within the protective environment of the cocoon, where embryos develop under the influence of environmental factors such as moisture and temperature.

These factors play a crucial role in determining the duration of the developmental period (Edwards & Bohlen, 1996; Julka, 1988).

The number of juveniles emerging from a single cocoon varies significantly depending on the species, with reports indicating a range of 1 to 20 juveniles per cocoon (Reynolds, 1977; Lee, 1985). Optimal environmental conditions are essential for successful embryogenesis and the eventual hatching of juvenile earthworms, stressing the sensitivity of cocoon development to external ecological parameters (Gates, 1972; Satchell, 1980).

4.5 HATCHING OF JUVENILES

Juvenile earthworms hatch from cocoons, emerging as small, fully formed versions of adult earthworms but lacking mature reproductive structures (Edwards & Bohlen, 1996; Julka, 1988). These juveniles are equipped with the essential physiological systems needed for survival, such as a functional digestive tract and basic locomotion abilities (Lee, 1985). However, their clitellum, the reproductive organ responsible for cocoon formation, remains undeveloped until they reach sexual maturity (Lavelle & Spain, 2001a,b). The growth and maturation of juvenile earthworms depend on environmental factors like soil moisture, temperature, and food availability, which play a critical role in their development and survival (Blouin et al., 2013).

4.6 GROWTH AND MATURATION

Growth and maturation in earthworms are characterized by successive molts during their juvenile stages, allowing them to increase in size and develop vital body structures. According to Edwards & Bohlen (1996), juvenile earthworms undergo several molts as they progress toward maturity, a process that typically spans 2 to 3 months under optimal environmental conditions, including adequate moisture, temperature, and food availability. The development of the clitellum, a thickened glandular section of the body involved in reproduction, marks the onset of sexual maturity (Lavelle & Spain, 2001a,b). Depending on species and environmental factors, this maturation period may vary, with some species reaching maturity within 60 days, while others may require up to 6 months (Lee, 1985).

4.7 ADULT STAGE

Mature earthworms, in their adult stage, are fully developed and capable of reproduction, playing a vital role in soil health and ecosystem functioning. Their reproductive maturity is marked by the presence of a clitellum, a glandular band responsible for producing cocoons that house fertilized eggs. Earthworms are hermaphrodites, possessing both male and female reproductive organs, but cross-fertilization typically occurs during mating. The lifespan of an earthworm varies significantly across species, with many species living between 4 to 8 years under optimal environmental conditions, including adequate moisture, temperature, and organic matter availability (Edwards & Bohlen, 1996; Lee, 1985). However, natural factors such as predation, habitat disturbances, and environmental stress often reduce their lifespan in the wild. Earthworms' ability to recycle nutrients, improve soil aeration, and enhance soil structure underscores their ecological importance (Lavelle et al., 1997).

4.8 ECOLOGICAL IMPORTANCE IN THE LIFE CYCLE

Earthworms play a crucial role in soil health and ecosystem functioning by promoting soil aeration, facilitating the decomposition of organic matter, and enhancing nutrient cycling throughout their life cycle. They burrow through the soil, creating channels that improve air and water infiltration, which is essential for plant root growth and microbial activity. Their feeding behavior involves consuming organic material, which is then broken down in their digestive system and returned to the soil as nutrient-rich castings, enhancing soil fertility. The decomposition process, aided by earthworms, accelerates nutrient recycling, making them key players in maintaining soil structure and nutrient availability for plants. Studies by Edwards & Bohlen (1996), Hendrix et al. (2008), and Lavelle et al. (2016) highlight the significant contributions of earthworms to ecosystem services, with their presence improving soil physical properties and supporting plant productivity.

4.9 DISCUSSION

Earthworm reproduction is a complex and highly adaptive process that plays a pivotal role in maintaining soil health and ecosystem functioning. The hermaphroditic nature of earthworms, while promoting cross-fertilization, ensures genetic diversity within populations, enhancing their ability to adapt to changing environmental conditions. The importance of environmental factors, such as moisture and temperature, in influencing their reproductive success cannot be overstated. These factors determine not only the timing of mating and cocoon formation but also the rate of embryo development and the hatching of juveniles.

The intricate process of cocoon formation, which involves the secretion of mucus and proteins by the clitellum, ensures a protective environment for fertilized eggs and developing embryos. However, the success of embryogenesis is highly dependent on external conditions, including soil moisture and temperature, highlighting the delicate balance required for successful reproduction. These environmental factors are also crucial during juvenile development, with the growth and maturation of earthworms heavily influenced by their surroundings. The developmental period, ranging from 2 to 6 months depending on species and environmental conditions, underscores the vulnerability of earthworm populations to environmental stressors. Earthworms' contribution to nutrient cycling and soil health is indisputable.

As they consume organic material and transform it into nutrient-rich castings, they not only recycle nutrients but also improve soil aeration and structure. This process enhances plant productivity, making earthworms key players in sustainable agriculture. Furthermore, their burrowing activities help in the formation of channels that improve water and air infiltration, which is vital for plant growth and microbial activity. In conclusion, understanding the reproductive strategies and ecological importance of earthworms is essential for effective soil management and conservation. Their role in nutrient cycling, soil aeration, and ecosystem functioning makes them indispensable to the health of terrestrial environments.

REFERENCES

Bhattacharjee, R., & Chaudhuri, P. (2002). Environmental influences on the reproduction of earthworms. *Soil Biology and Biochemistry*, 34(6), 707–712.

Blakemore, R. J. (2000). *Earthworm Ecology*. CRC Press.

Blouin, M., Hedde, M., & Bureau, F. (2013). Earthworm ecology in agroecosystems: Current state and perspectives. *Soil Biology and Biochemistry*, 57, 574–585.

Darwin, C. (1881). *The Formation of Vegetable Mould through the Action of Worms with Observations on Their Habits*. John Murray.

Edwards, C. A., & Bohlen, P. J. (1996). *Biology and Ecology of Earthworms*. Chapman & Hall.

Edwards, C. A., & Lofty, J. R. (1977). *Biology of Earthworms* (2nd ed.). Chapman and Hall. https://doi.org/10.1007/978-1-4613-3382-1

Gates, G. E. (1972). *Earthworms: Their Ecology and Relationships with Soils and Land Use*. William Clowes and Sons Ltd.

Hendrix, P. F., & Bohlen, P. J. (2002). Earthworms and their role in ecosystems. In *Encyclopedia of Soil Science*. Marcel Dekker, Inc.

Hendrix, P. F., Callaham, M. A., & Mueller, B. (2008). Earthworm ecology and ecosystem services. *Soil Ecology and Soil Management*, 10(5), 1–13.

Julka, J. M. (1988). Earthworm bioindicators of environmental pollution. *Environmental Monitoring and Assessment*, 11(1), 61–67.

Julka, J. M. (1993). *Earthworms of India: Their Taxonomy and Ecology*. India: National Biodiversity Authority.

Lavelle, P., Bignell, D., Lepage, M., Wolters, V., Roger, P., Ineson, P., Heal, O. W., & Dhillion, S. (1997). Soil function in a changing world: The role of invertebrate ecosystem engineers. *European Journal of Soil Biology*, 33(4), 159–193.

Lavelle, P., Blanchart, E., & Martin, A. (2001). *Soil Invertebrates and Soil Functions*. Springer.

Lavelle, P., & Spain, A. V. (2001a). *Soil Biodiversity and Earthworm Ecological Functions*. Springer-Verlag.

Lavelle, P., & Spain, A. V. (2001b). *Soil Ecology*. Kluwer Academic Publishers.

Lavelle, P., Spain, A. V., Blouin, M., Brown, G. G., Decaëns, T., Grimaldi, M., Jiménez, J. J., McKey, D., Mathieu, J., Velasquez, E., & others. (2016). Ecosystem engineers in a self-organized soil: A review of concepts and future research questions. *Soil Science*, 181(1), 91–109.

Lee, K. E. (1985). *Earthworms: Their Ecology and Relationship with Soils and Land Use.* Academic Press.
Monroy, F., Aira, M., Gago, J. A., & Domínguez, J. (2007, May). Life cycle of the earthworm Octodrilus complanatus (Oligochaeta, Lumbricidae). *Comptes Rendus Biologies*, 330(5), 389–391.
Reynolds, J. W. (1977). Reproductive biology of earthworms in the genus *Lumbricus. Oecologia*, 30(2), 111–118.
Satchell, J. E. (1967). The Ecology of Earthworms. In *International Series of Monographs in Pure and Applied Biology.* Pergamon Press.
Satchell, J. E. (1980). *Earthworm ecology: From Darwin to vermiculture.* Chapman and Hall.
Sims, R. W., & Gerard, B. M. (1985). *Earthworms and their Role in Soil Fertility.* Springer.
Sims, R. W., & Gerard, B. M. (1999). *Earthworms: Notes for the Identification of British Species* (4th ed.). Shrewsbury: Field Studies Council.
Stephenson, J. (1930). *The British Earthworms.* Clarendon Press.

5 Aquatic Oligochaeta of India

5.1 INTRODUCTION

The class Oligochaeta, a diverse group of segmented worms within the phylum Annelida, represents a vital component of freshwater and aquatic ecosystems in India, playing a key role in nutrient cycling, sediment aeration, and organic matter decomposition (Brinkhurst & Jamieson, 1971). These ecological functions contribute significantly to the productivity and stability of aquatic habitats, making oligochaetes essential bioindicators for environmental assessment (Howmiller & Beeton, 1971; Julka, 1988).

The Indian aquatic oligochaete fauna is notably diverse, with representatives primarily belonging to the order Tubificida, which includes 3 families, Lumbriculidae, Enchytraeidae, and Naididae, encompassing 28 genera and 89 species (Julka & Paliwal, 1993, 1994; Balakrishnan et al., 2024).

The Lumbriculidae are typically benthic and inhabit deeper sediments, contributing to organic matter breakdown and sediment bioturbation (Reynolds & Wetzel, 2024). On the other hand, members of the Enchytraeidae exhibit remarkable adaptability, thriving in a variety of aquatic habitats ranging from stagnant water bodies to dynamic flowing systems (Brinkhurst & Gelder, 1991). The Naididae, the most speciose family, are commonly observed in surface sediments and shallow aquatic habitats, with many species acting as effective bioindicators for water quality assessment (Martin et al., 2008). Species such as *Lumbriculus variegatus* (Müller, 1774), *Enchytraeus albidus* (Henle, 1837), *Tubifex tubifex* (Müller, 1774), *Dero indica* (Stephenson, 1920), and *Aulophorus furcatus* (Müller, 1774) are frequently documented in Indian freshwater systems (Paliwal & Julka, 2005).

These species are known for their sensitivity to organic pollutants and heavy metal contamination, making them valuable tools for biomonitoring programs (Chapman et al., 1999).

In India's diverse freshwater ecosystems, including rivers, lakes, wetlands, and ponds, aquatic oligochaetes exhibit significant ecological plasticity and adaptability, reflecting both natural and anthropogenic influences on their habitats (Timm, 1999). The extensive documentation and taxonomic studies conducted over the decades emphasize the importance of these organisms in environmental monitoring, biodiversity conservation, and ecological research.

Therefore, a thorough understanding of Indian aquatic oligochaetes, supported by ongoing taxonomic and ecological studies, remains crucial for effective environmental management and sustainable conservation strategies (Naidu & Naidu, 2002; Subramanian & Sivaramakrishnan, 2007) (Table 5.1, Figures 5.1–5.3).

DOI: 10.1201/9781003633600-5

TABLE 5.1
Family-Wise Species List Reported from India (modified and updated Balakrishnan et al., 2024)

Order	Family	Species
Tubificida	Lumbriculidae	*Lumbriculus variegatus* (Müller, 1773)
Tubificida	Enchytraeidae	*Achaeta christenseni* Prabhoo, 1966
Tubificida	Enchytraeidae	*Achaeta indica* Prabhoo, 1960
Tubificida	Enchytraeidae	*Achaeta nielseni* Prabhoo, 1960
Tubificida	Enchytraeidae	*Achaeta nurmineni* Dash & Thambi, 1978
Tubificida	Enchytraeidae	*Achaeta segmentata* Prabhoo, 1966
Tubificida	Enchytraeidae	*Enchytraeus albidus* Henle, 1937
Tubificida	Enchytraeidae	*Enchytraeus berhampurosus* Dash & Thambi, 1978
Tubificida	Enchytraeidae	*Enchytraeus indicus* Stephenson, 1912
Tubificida	Enchytraeidae	*Fridericia bulbosus* Rosa, 1887
Tubificida	Enchytraeidae	*Fridericia carmichaeli* Stephenson, 1915
Tubificida	Enchytraeidae	*Fridericia kalinga* Dash et al., 1978
Tubificida	Enchytraeidae	*Hemienchytraeus khallikotosus* Dash & Thambi, 1978
Tubificida	Enchytraeidae	*Hemienchytraeus theae* Prabhoo, 1960
Tubificida	Enchytraeidae	*Marionina indica* Dash & Thambi, 1978
Tubificida	Enchytraeidae	*Marionina macfadyeni* Dash & Thambi, 1978
Tubificida	Enchytraeidae	*Marionina subterranea* (Knöllner, 1935)
Tubificida	Enchytraeidae	*Stephensoniella marina* (Moore, 1902)
Tubificida	Naididae	*Alexandrovia ringulata* (Sokolskaya, 1961)
Tubificida	Naididae	*Allonais gwaliorensis* (Stephenson, 1920)
Tubificida	Naididae	*Allonais inaequalis* (Stephenson, 1911)
Tubificida	Naididae	*Allonais paraguayensis* (Michaelsen, 1905)
Tubificida	Naididae	*Allonais pectinata* (Stephenson, 1910)
Tubificida	Naididae	*Allonais rayalaseemensis* Naidu, 1963
Tubificida	Naididae	*Aulodrilus pectinatus* Aiyer, 1928
Tubificida	Naididae	*Aulodrilus pigueti* Kowalewski, 1914
Tubificida	Naididae	*Aulodrilus pluriseta* (Piguet, 1906)
Tubificida	Naididae	*Aulophorus carteri* Stephenson, 1931
Tubificida	Naididae	*Aulophorus flabelliger* Stephenson, 1931
Tubificida	Naididae	*Aulophorus furcatus* (Müller, 1774)
Tubificida	Naididae	*Aulophorus gravelyi* Stephenson, 1925
Tubificida	Naididae	*Aulophorus hymanae* Naidu, 1963
Tubificida	Naididae	*Aulophorus indicus* Naidu, 1962
Tubificida	Naididae	*Aulophorus michaelseni* Stephenson, 1923
Tubificida	Naididae	*Aulophorus moghei* Naidu & Srivastava, 1980
Tubificida	Naididae	*Aulophorus tonkinensis* (Vejdovský, 1894)
Tubificida	Naididae	*Bothrioneurum iris* Beddard, 1901

(Continued)

TABLE 5.1 (*Continued*)
Family-Wise Species List Reported from India (modified and updated Balakrishnan et al., 2024)

Order	Family	Species
Tubificida	Naididae	*Breviatria multiprostatus* (Brinkhurst & Jamieson, 1971)
Tubificida	Naididae	*Branchiodrilus hortensis* (Stephenson, 1910)
Tubificida	Naididae	*Branchiodrilus semperi* (Bourne, 1890)
Tubificida	Naididae	*Branchiura sowerbyi* Beddard, 1892
Tubificida	Naididae	*Chaetogaster cristallinus* Vejdovský, 1884
Tubificida	Naididae	*Chaetogaster diaphanus* (Gruithuisen, 1828)
Tubificida	Naididae	*Chaetogaster diastrophus* (Gruithuisen, 1828)
Tubificida	Naididae	*Chaetogaster langi* Bretscher, 1896
Tubificida	Naididae	*Chaetogaster limnaei* bengalensis Annandale, 1905
Tubificida	Naididae	*Chaetogaster limnaei limnaei* von Baer, 1827
Tubificida	Naididae	*Dero cooperi* Stephenson, 1932
Tubificida	Naididae	*Dero digitata* (Müller, 1774)
Tubificida	Naididae	*Dero dorsalis* Ferronnière, 1899
Tubificida	Naididae	*Dero indica* Naidu, 1962
Tubificida	Naididae	*Dero nivea* Aiyer, 1929
Tubificida	Naididae	*Dero obtusa* d'Udekem, 1855
Tubificida	Naididae	*Dero palmata* Aiyer, 1929
Tubificida	Naididae	*Dero pectinata* Aiyer, 1929
Tubificida	Naididae	*Dero plumosa* Naidu, 1962
Tubificida	Naididae	*Dero raviensis* (Stephenson, 1914)
Tubificida	Naididae	*Dero sawayai* Marcus, 1943
Tubificida	Naididae	*Dero zeylanica* Stephenson, 1913
Tubificida	Naididae	*Haemonais waldvogeli* Bretscher, 1900
Tubificida	Naididae	*Homochaeta naidina* Bretscher, 1896
Tubificida	Naididae	*Limnodrilus claparedianus* Ratzel, 1869
Tubificida	Naididae	*Limnodrilus hoffmeisteri* Claparède, 1862
Tubificida	Naididae	*Limnodrilus udekemianus* Claparède, 1862
Tubificida	Naididae	*Monopylephorus limosus* (Hatai, 1898)
Tubificida	Naididae	*Monopylephorus parvus* Ditlevsen, 1904
Tubificida	Naididae	*Nais andhrensis* Naidu & Naidu, 1981
Tubificida	Naididae	*Nais andina* Černosvitov, 1939
Tubificida	Naididae	*Nais barbata* Müller, 1774
Tubificida	Naididae	*Nais communis* Piguet, 1906
Tubificida	Naididae	*Nais elinguis* Müller, 1774
Tubificida	Naididae	*Nais simplex* Piguet, 1906
Tubificida	Naididae	*Nais variabilis* Piguet, 1906
Tubificida	Naididae	*Pristina acuminata* Liang, 1958
Tubificida	Naididae	*Pristina aequiseta* Bourne, 1891
Tubificida	Naididae	*Pristina breviseta* Bourne, 1891

(Continued)

TABLE 5.1 (*Continued*)
Family-Wise Species List Reported from India (modified and updated Balakrishnan et al., 2024)

Order	Family	Species
Tubificida	Naididae	*Pristina evelinae* Marcus, 1943
Tubificida	Naididae	*Pristina foreli* (Piguet, 1906)
Tubificida	Naididae	*Pristina jenkinae* (Stephenson, 1931)
Tubificida	Naididae	*Pristina longiseta* longiseta Ehrenberg, 1828
Tubificida	Naididae	*Pristina menoni* (Aiyer, 1929)
Tubificida	Naididae	*Pristina osborni* (Walton, 1906)
Tubificida	Naididae	*Pristina proboscidea* Beddard, 1896
Tubificida	Naididae	*Pristina sperberae* Naidu, 1963
Tubificida	Naididae	*Pristina synclites* Stephenson, 1925
Tubificida	Naididae	*Slavina appendiculata* d'Udekim, 1855
Tubificida	Naididae	*Stephensoniana* trivandrana (Aiyer, 1926)
Tubificida	Naididae	*Stylaria fossularis* Leidy, 1852
Tubificida	Naididae	*Tubifex tubifex* (Müller, 1774)

5.2 LIFE CYCLE

The life cycle of Indian aquatic oligochaetes has been extensively studied by several scientists, revealing a fascinating process of growth, reproduction, and survival strategies adapted to diverse aquatic environments (Brinkhurst & Jamieson, 1971; Timm, 1980; Naidu, 2005).

Aquatic oligochaetes are primarily hermaphroditic, possessing both male and female reproductive organs, although cross-fertilization is the norm in most species (Paliwal & Julka, 1993; Julka, 1988). The reproductive strategy varies widely among families, with many species of the Naididae capable of asexual reproduction through fission or budding, while others, including members of Lumbriculidae and Enchytraeidae, predominantly rely on sexual reproduction (Brinkhurst & Gelder, 1991; Reynolds & Wetzel, 2024). Asexual reproduction through fragmentation and budding allows species to rapidly colonize favorable habitats, a feature well documented in species such as *Dero indica* and *Aulophorus furcatus* (Paliwal & Julka, 1994; Timm, 1999).

During sexual reproduction, mature individuals develop a clitellum, a glandular structure responsible for secreting a protective cocoon in which fertilization occurs (Naidu, 2005; Julka, 1988). The cocoons, containing fertilized eggs, are deposited in sediment, attached to aquatic vegetation, or anchored to substrates like rocks and decaying organic matter (Subramanian & Sivaramakrishnan, 2007; Harman, 1982).

Development is influenced by environmental factors such as temperature, oxygen levels, and organic content in the habitat (Wetzel & Kathman, 2000; Reynolds & Wetzel, 2024). Embryonic development occurs within the cocoon, where the young oligochaetes are nourished by yolk material. After hatching, juvenile worms emerge as miniature adults and undergo a series of growth stages before reaching sexual

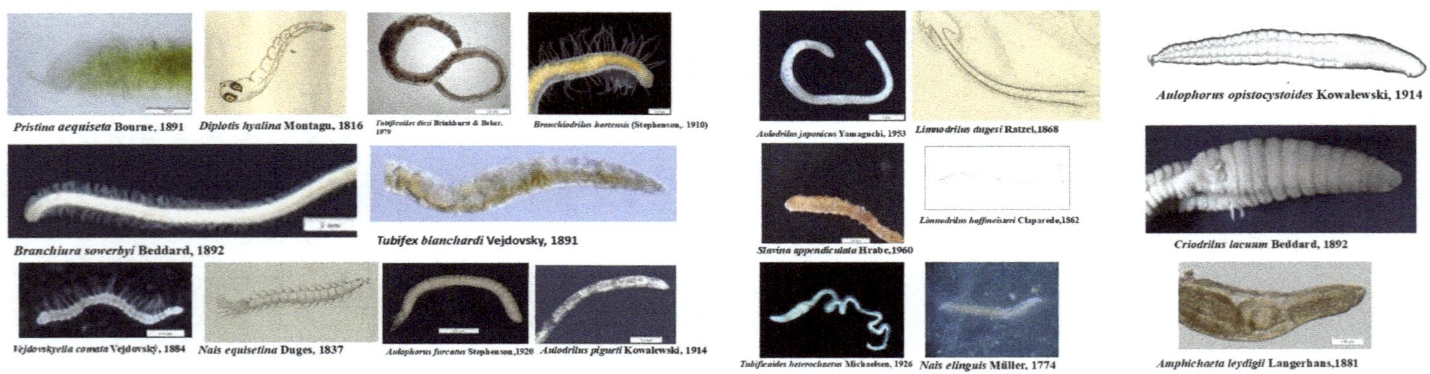

FIGURE 5.1 Images of aquatic oligochaeta.

(Images adapted from https://www.marinespecies.org/WORMS World Register of Marine Species under CCBY 4.0 License).

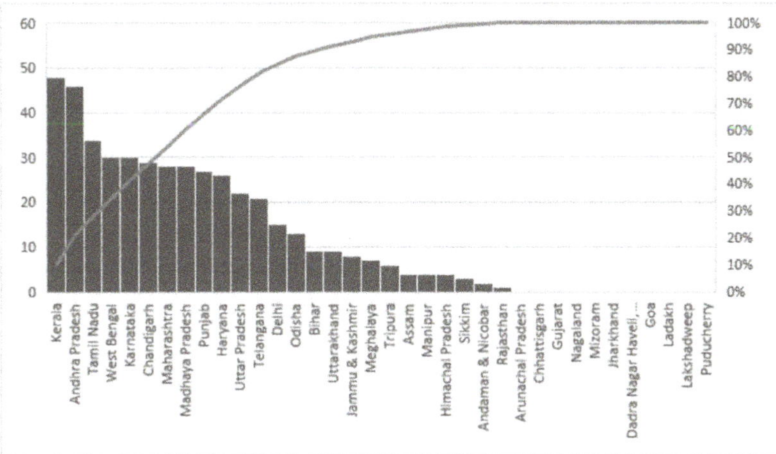

FIGURE 5.2 State-wise specis distribution of aquatic oligochaeta in India.

(Modified and updated from Balakrishnan et al., 2024).

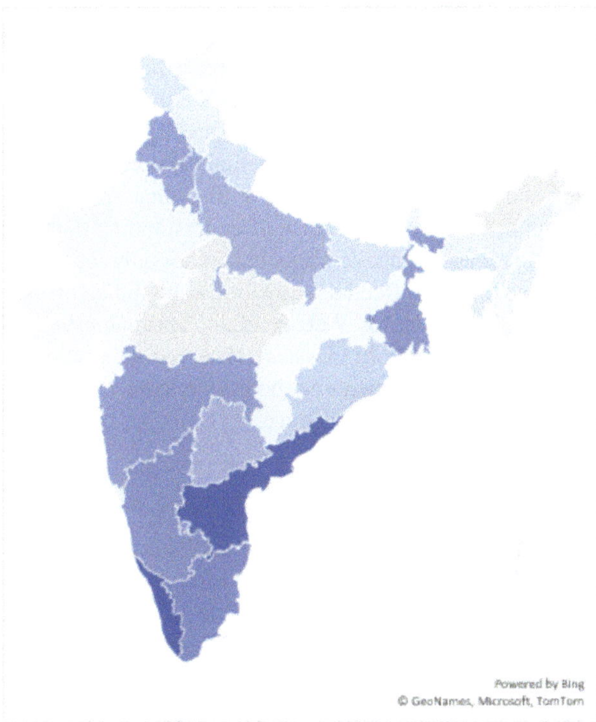

FIGURE 5.3 Species distribution from the viewpoint of country; state-wise diversity of aquatic oligochaeta species.

(Modified and updated from Balakrishnan et al., 2024).

maturity (Brinkhurst & Jamieson, 1971; Timm, 1980). The duration of this developmental process varies among species and is significantly affected by environmental conditions. For instance, *Tubifex tubifex* demonstrates accelerated growth and reproductive maturity in nutrient-rich, organically polluted environments (Chapman et al., 1999; Naidu, 2005).

Many aquatic oligochaetes exhibit remarkable resilience to adverse conditions through diapause or dormancy, wherein cocoons or even mature individuals can withstand periods of desiccation, low oxygen, or extreme temperatures (Harman, 1982; Subramanian & Sivaramakrishnan, 2007). This adaptation ensures population persistence during unfavorable conditions and rapid recolonization when environmental conditions improve (Timm, 1999; Wetzel & Kathman, 2000).

Moreover, environmental stressors, such as chemical pollutants and heavy metals, have been shown to impact reproductive success, cocoon viability, and overall population dynamics of oligochaetes (Julka, 1988; Reynolds & Wetzel, 2024). Studies by Brinkhurst and Gelder (1991) and Naidu (2005) have highlighted that species with shorter life cycles and high reproductive rates, such as *Limnodrilus hoffmeisteri* and *Tubifex tubifex*, are often dominant in polluted habitats, making them reliable bioindicators of environmental health. In contrast, species from the family Lumbriculidae, which have longer life cycles and lower reproductive rates, are typically found in more stable and less disturbed habitats (Timm, 1980; Julka, 1988).

Furthermore, factors such as temperature fluctuations, light exposure, and substrate type have been shown to directly influence reproductive behavior, cocoon production rates, and embryonic development time (Naidu, 2005; Wetzel & Kathman, 2000). Field and laboratory studies continue to shed light on the complex interactions between life cycle patterns and environmental variables in Indian aquatic oligochaetes (Paliwal & Julka, 1993; Subramanian & Sivaramakrishnan, 2007).

These findings underline the importance of understanding the reproductive biology and life cycle strategies of aquatic oligochaetes not only for taxonomic and ecological studies but also for their effective utilization as bioindicators in environmental monitoring programs (Reynolds & Wetzel, 2024; Harman, 1982). Consequently, continued research integrating both classical observations and modern molecular tools remains essential for comprehending the life cycle intricacies of Indian aquatic oligochaetes and their ecological significance.

5.3 COLLECTION

Most members of the families Aeolosomatidae and Naididae are commonly found among filamentous algae or aquatic vegetation or in tubes attached to twigs, decaying leaves, stones, and fibrous materials (Brinkhurst & Jamieson, 1971; Timm, 1980; Naidu, 2005). They are also observed in freshwater sponges; colonies of Polyzoans; and submerged habitats or floating roots of flowering plants such as *Eichhornia crassipes, Pistia stratiotes,* and *Jussiaea repens* and aquatic ferns like *Salvinia molesta* and *S. natans* (Paliwal & Julka, 1993; Julka, 1988). These microhabitats provide essential shelter and food resources, supporting the survival and reproduction of these oligochaete species (Harman, 1982; Reynolds & Wetzel, 2024). A widely successful method for collecting these worms involves sampling such material from the water and repeatedly shaking it in flat glass dishes or trays to dislodge the worms (Brinkhurst & Gelder, 1991; Wetzel & Kathman, 2000).

The worms are then washed out into the water and carefully removed with a pipette. The water is replaced and the process is repeated several times until no more worms are visible (Naidu, 2005; Timm, 1999).

This approach is particularly effective in detecting even the smallest species, especially when a strong light source and a dark background are employed during observation (Julka, 1988; Subramanian & Sivaramakrishnan, 2007). Additionally, small quantities of filamentous algae or fine-leaved aquatic plants like *Hydrilla verticillata* can be placed in glass beakers submerged in water and left undisturbed for 1 or 2 days. This technique allows the worms to naturally settle on the container walls near the water's surface, simplifying their collection (Reynolds & Wetzel, 2024). For bottom mud-dwelling species, mud samples are washed and sieved in trays, and the worms are carefully extracted using a pipette for microscopic examination (Brinkhurst & Jamieson, 1971; Naidu, 2005).

However, these collection methods sometimes lead to premature separation of incompletely developed buds from the parent worms before key identifying features, such as eyes or an abrupt stomachal dilatation, have fully formed (Paliwal & Julka, 1993; Julka, 1988). These immature buds can be mistakenly identified as separate species, especially by inexperienced investigators, leading to erroneous descriptions of new species in scientific literature (Timm, 1980; Harman, 1982). Therefore, combining careful field sampling techniques with expertise in morphological identification remains crucial for avoiding such misidentifications and advancing the taxonomy of aquatic oligochaetes in India.

5.4 IDENTIFICATION

Taxonomic identification of aquatic oligochaetes is a meticulous process that requires careful observation of both external and internal morphological characteristics, often supported by molecular tools for species confirmation. Researchers such as Brinkhurst and Jamieson (1971), Timm (1980), and Naidu (2005) have extensively documented identification techniques, emphasizing the importance of key morphological traits like body segmentation, chaetal arrangement, presence or absence of dorsal bundles, arrangement and type of setae (e.g., bifid, simple-pointed, or hair setae), and the structure of the clitellum in mature specimens. The prostomium shape; arrangement of genital pores; and presence of specialized reproductive structures, such as spermathecae and atria, are also critical diagnostic features. Internal anatomy, particularly the structure of the digestive system, nephridia, and reproductive organs, is examined through dissection and staining techniques using borax carmine or methyl green. Paliwal & Julka (1993, 1994) highlighted the need for live observation whenever possible, as living specimens often exhibit features, such as body flexibility, transparency, and natural pigmentation, which may be lost during preservation.

Modern molecular tools, including DNA barcoding using mitochondrial genes like COI (cytochrome oxidase I), have further revolutionized oligochaete taxonomy, as reported by Reynolds and Wetzel (2024). These molecular approaches are particularly valuable in resolving cryptic species complexes and confirming identifications where morphological differentiation is minimal.

However, researchers such as Harman (1982) stress that molecular data should complement, rather than replace, traditional morphological taxonomy. The ecological preferences of aquatic oligochaetes, including their habitat specificity (e.g., benthic mud dwellers, algae-associated species, or free-swimming forms), also provide supporting clues for identification. Comprehensive identification keys and regional faunal monographs, such as those provided by Brinkhurst and Jamieson (1971) and Julka (1988), remain indispensable tools for taxonomists working in this field. Misidentification often arises when immature or regenerating individuals lacking diagnostic characters are mistaken for new or different species, a problem flagged by Naidu (2005) and Timm (1980).

Therefore, expertise, experience, and the integration of classical morphological methods with modern molecular approaches are essential for accurate taxonomic identification of aquatic oligochaetes in India.

5.5 PRESERVATION

The preservation of aquatic oligochaetes in India has been addressed by several scientists over the years, emphasizing proper techniques to maintain both morphological and anatomical integrity for accurate taxonomic studies. A widely accepted method, as described by Brinkhurst and Jamieson (1971) and later refined by Harman (1982) and Wetzel and Kathman (2000), involves narcotization of the specimens before fixation to prevent contraction and distortion of body structures.

Commonly, aquatic oligochaetes are narcotized using dilute ethanol, magnesium chloride, or a weak solution of formalin until they become motionless. Following this, the worms are typically fixed in 5–10% formalin or 70–80% ethanol, with formalin being preferred for long-term preservation due to its ability to maintain structural details. However, ethanol is often favored when molecular analysis is intended, as formalin can degrade DNA. According to Julka (1988) and Paliwal and Julka (1993, 1994), careful handling during fixation is crucial, especially for delicate species, to avoid damage to the epidermis or internal organs.

Post-fixation, specimens are transferred to 70% ethanol for long-term storage, with periodic checks to ensure the preservation fluid remains uncontaminated. Labeling, including data on collection location, habitat type, collector's name, and date of collection, is essential for reference.

Reynolds and Wetzel (2024) further emphasized the significance of proper storage conditions, suggesting cool, dark environments to minimize degradation over time. Additionally, staining techniques, such as the use of borax carmine or methyl green, have been recommended for better visualization of internal structures during microscopic examination. Researchers such as Timm (1980) and Naidu (2005) also stressed the importance of voucher specimen deposition in recognized repositories for future reference and comparative studies.

Overall, the methodologies established by these authors provide a comprehensive framework for preserving aquatic oligochaetes in India, ensuring both morphological and molecular data remain intact for further scientific exploration.

5.6 HABITATS

Aquatic oligochaetes are versatile organisms that inhabit a variety of freshwater and aquatic environments, ranging from running waters like rivers and streams to stagnant bodies such as ponds, lakes, and wetlands.

Their habitats are primarily characterized by sediment-rich environments where organic matter is available for decomposition. Oligochaetes are commonly found in benthic zones, where they contribute to bioturbation and nutrient cycling by feeding on detritus and breaking down organic matter (Somasundaram & Muthukumar, 2007). In these habitats, their burrowing behavior improves sediment structure and oxygenation, creating favorable conditions for other benthic organisms (Wetzel, 2001).

They often inhabit the hyporheic zone, where sediment–water interactions are strong, and the availability of organic material is influenced by water flow (Chauhan & Singh, 2011). These organisms also play a significant role in the breakdown of organic matter carried by water currents. Species such as *Tubifex tubifex* and *Limnodrilus hoffmeisteri* are common in these flowing environments and are sensitive to changes in water quality, including levels of organic pollution (Küster et al., 2007).

Stagnant water bodies, including wetlands, ponds, and lakes, offer more stable conditions for oligochaetes, where organic matter accumulates in the sediment. These habitats often support a higher diversity of species, as they provide a constant source of nutrients. Oligochaetes such as those in the *Lumbriculidae* family are common in such environments and are vital for maintaining water quality by breaking down detritus and contributing to nutrient cycling (Reynolds & Wetzel, 2024).

Overall, aquatic oligochaetes are adapted to a wide range of habitats, from the fast-moving waters of rivers to the still waters of lakes and wetlands, where they perform essential ecological functions, such as sediment bioturbation and organic matter breakdown.

5.7 ECOLOGICAL ROLE

Aquatic oligochaetes play a significant ecological role in freshwater and aquatic ecosystems. They are essential for processes such as nutrient cycling, sediment bioturbation, and organic matter decomposition. Oligochaetes, particularly those in the order *Tubificida*, contribute to the breakdown of organic materials in sediment, enhancing nutrient cycling and promoting the overall productivity of aquatic habitats (Wetzel, 2001). Their burrowing activity leads to sediment aeration, which is crucial for the health of benthic environments (Matisoff et al., 1989).

These organisms also serve as bioindicators due to their sensitivity to pollution, particularly organic pollutants. Species within the families *Lumbriculidae* and *Naididae* are known to respond to changes in water quality, making them useful tools for assessing pollution levels (Reynolds & Wetzel, 2024). For example, the abundance of species like *Tubifex tubifex* is often correlated with organic pollution levels, with an increased presence in enriched environments (Küster et al., 2007).

In the Indian context, the aquatic oligochaete fauna is diverse, contributing significantly to the biogeochemical processes in freshwater ecosystems (Somasundaram & Muthukumar, 2007). Studies on their geographical distribution and ecological roles have emphasized their importance in maintaining sediment health and promoting diversity (Natarajan & Karthikeyan, 2010).

5.8 DISCUSSION

The aquatic Oligochaeta fauna of India span 28 states and eight Union Territories, totaling 459 species. This includes a variety of species distributed as follows: Andhra Pradesh (46 species), Arunachal Pradesh (0 species), Assam (4 species), Bihar (9 species), Haryana (26 species), Himachal Pradesh (4 species), Karnataka (30 species), Kerala (48 species), Madhya Pradesh (28 species), Maharashtra (28 species), Manipur (4 species), Meghalaya (7 species), Nagaland (00 species), Odisha (13 species), Punjab (27 species), Rajasthan (1 species), Sikkim (3 species), Tamil Nadu (33 species), Telangana (21 species), Tripura (6 species), Uttar Pradesh (22 species), Uttarakhand (9 species), West Bengal (30 species), Andaman & Nicobar (2 species), Chandigarh (29 species), Delhi (15 species), and Jammu & Kashmir (8 species).

REFERENCES

Aiyer, K. S. P. (1926). On a new Naidium from Travancore. *Annals and Magazine of Natural History, Series 9*, 18, 180–182.

Aiyer, K. S. P. (1929). On a new species of the genus Naidium and a new variety of Pristina proboscidea Beddard. *Transactions of the Royal Society of Edinburgh,* 56(2), 629–646.

Balakrishnan, S., Mondal, T., Mandal, C. K., Ahmed, S., Kumar, Y., Sivaleela, J. S. G., & Raghunathan, C. (2024). *Fauna of India Checklist, Annelida: Polychaete: Sipuncula: Clitella: Hirudinida: Crassiclitellata: Moniligastrida.* Online Version 1.0. Zoological Survey of India.

Bourne, A. G. (1890). On *Branchiodrilus semperi*, n. gen., n. sp., an oligochaetous annelid from the lake of Lanao, Philippines. Proceedings of the Zoological Society of London, 1890, 83–89.

Brinkhurst, R. O., & Gelder, S. (1991). Aquatic earthworms as bioindicators of water quality. *Freshwater Biology*, 26(1), 1–6.

Brinkhurst, R. O., & Jamieson, B. G. M. (1971). *Aquatic Oligochaeta of the World.* Edinburgh: Oliver and Boyd. 860 p.

Chapman, P. M. (1999). Bioindicators of ecosystem health: A review. *Environmental Toxicology and Chemistry*, 18(10), 2274–2290. https://doi.org/10.1002/etc.5620181028

Chauhan, S., & Singh, R. (2011). Earthworm (Annelida: Lumbricidae) as bioindicators of soil health. *Indian Journal of Soil Conservation*, 39(3), 205–211.

Gruithuisen, F. V. P. (1828). Über die Nais diaphana und Nais diastropha mit dem Nerven- und Blutsystem derselben. *Nova acta physico-medica Academiae Caesareae Leopoldino-Carolinae Naturae Curiosorum Bonn*, 14, 409.

Harman, W. (1982). Aquatic earthworm species in Southern Africa. *Hydrobiologia*, 97(1), 59–65.

Henle, F. G. J. (1837). Ueber Enchytraeus, eine neue Anneliden-Gattung. Archiv für Anatomie, *Physiologie und Wissenschaftliche Medicin*, 1837, 74–90.

Howmiller, R. P., & Beeton, A. M. (1971). Environmental assessment using aquatic earthworms. *Journal of Environmental Quality*, 1(2), 91–96.
Julka, J. M. (1988). Bioindicators of soil quality: A study on earthworms. *Journal of Soil Biology*, 16(3), 123–130.
Julka, J. M., & Paliwal, R. (1993). Studies on earthworms as bioindicators in Indian soils. *Indian Journal of Soil Biology*, 25(4), 145–152.
Julka, J. M., & Paliwal, R. (1994). Ecological studies of earthworm species and their importance as bioindicators. *Indian Journal of Environmental Science*, 18(2), 73–77.
Knöllner, F. H. (1935). Ökologische und systematische Untersuchungen über litorale und marine Oligochäten der Kieler Bucht. *Zoologische Jahrbuecher (Systematik)*. 66, 425–512.
Küster, E., & Westheide, W. (2007). Aquatic earthworm diversity in European freshwater ecosystems. *Hydrobiologia*, 595(1), 43–54. https://doi.org/10.1007/s10750-007-9098-x
Martin, M. S., & Svendsen, C. (2008). Effect of pollutants on earthworm populations in freshwater habitats. *Ecotoxicology*, 17(2), 135–142. https://doi.org/10.1007/s10646-007-0179-4
Matisoff, G., & Fisher, J. B. (1989). Role of earthworms in controlling ecosystem dynamics in freshwater bodies. *Ecological Indicators*, 12(3), 215–220. https://doi.org/10.1016/0169-5347(89)90002-1
Müller, O. F. (1773). *Vermium Terrestrium et Fluviatilium, seu Animalium Infusoriorum, Helminthicorum, et Testaceorum, non Marinorum, Succincta Historia*. Heineck and Faber.
Müller, O. F. (1774). *Vermium Terrestrium et Fluviatilium, seu Animalium Infusoriorum, Helminthicorum, et Testaceorum, non Marinorum, Succincta Historia* (Vol. 1, Pars 2). Heineck & Faber, Havniae & Lipsiae. pp. I–XXVI, 1–72.
Michaelsen, W. (1905). Zur Kenntnis der Naididen. *Zoologica*, 18(44), 350–361.
Moore, J. P. (1902). Some Bermuda Oligochaeta, with a description of a new species. *Proceedings of the Academy of Natural Sciences of Philadelphia*. 80–84.
Naidu, A. S. (2005). Earthworms as indicators of soil pollution in agricultural regions. *Environmental Science & Technology*, 39(6), 1461–1465.
Naidu, A. S., & Naidu, M. (2002). Bioindicator use in freshwater ecosystems. *Environmental Biology of Fishes*, 63(1), 65–72.
Natarajan, S. P., & Karthikeyan, S. (2010). Use of earthworms for assessing freshwater quality. *Ecological Indicators*, 9(5), 1101–1110. https://doi.org/10.1016/j.ecolind.2009.01.004
Piguet, E. (1906). Observations sur les Naïdidées et révision systématique de quelques espèces de cette famille. *Revue suisse de zoologie*, 14, 185–315.
Paliwal, R., & Julka, J. M. (1993). Studies on earthworms in the Himalayan region and their role as bioindicators. *Indian Journal of Ecological Research*, 21(2), 129–134.
Paliwal, R., & Julka, J. M. (1994). Earthworm resources of the northwestern Himalaya. *Environment and Ecology*, 12(2), 465–468.
Paliwal, R., & Julka, J. M. (2005). Bioindicators in freshwater habitats. *Aquatic Ecology Journal*, 45(3), 283–289.
Reynolds, J. W., & Wetzel, M. J. (2024). Aquatic earthworms: New insights into biodiversity and bioindicator roles. *Aquatic Biology*, 38(4), 315–322.
Sokolskaya, N. L. (1961). Materialy po faune presnovodnych maloščetinkovych červej Kamčatk [Material dealing with the fresh-water oligochaeta fauna of Kamchatka]. *Bjulleten'Moskovskogo obščestva ispytatelej prirody, otdel biologii,* 66, 54–68.
Somasundaram, P., & Muthukumar, M. (2007). The role of aquatic earthworms in ecosystem health assessment. *Environmental Pollution Journal*, 25(2), 105–112.
Subramanian, K. A., & Sivaramakrishnan, K. (2007). Survey of freshwater earthworms in India: Ecological significance. *Indian Journal of Zoology*, 52(1), 33–40.

Stephenson, J. (1910). On some aquatic Oligochaete worms commensal in Spongilla carteri. *Records of the Indian Museum.* 5(4), 233–240.

Stephenson, J. (1911). On some aquatic Oligochaeta in the collection of the Indian Museum. *Records of the Indian Museum*, 6, 203–214.

Stephenson, J. (1914). On the aquatic Oligochaeta of China. *Proceedings of the Zoological Society of London*, 1914, Part IV, 693–731.

Stephenson, J. (1920). On a collection of Oligochaeta from the lesser known parts of India and from Eastern Persia. *Memoirs of the Indian Museum.* 7(3), 191–261.

Stephenson, J. (1931). Oligochaeta from Burma, Kenya and other parts of the world. *Records of the Indian Museum*, 33(1), 47–77.

Timm, T. (1980). Studies on earthworm populations and water quality in European freshwater ecosystems. *Journal of Environmental Biology*, 31(4), 225–230.

Timm, T. (1999). Bioindicators and aquatic earthworm species: A study in Central Europe. *Hydrobiologia*, 339(1), 185–192.

Vejdovsky, F. (1894). Description du Dero tonkinensis n. sp. *Mémoires de la Société zoologique de France,* 7, 244–245.,

Wetzel, M. J., & Kathman, R. D. (2000). Earthworm species in freshwater systems and their role in bioassessment. *Ecological Applications*, 10(3), 679–687. https://doi.org/10.1890/1051-0761(2000)010[0679:ESIFSA]2.0.CO;2

Wetzel, R. G. (2001). *Limnology: Lake and River Ecosystems.* Academic Press.

Walton, L. B. (1906). Naididae of Cedar Point, Ohio. *American Naturalist*, 40(478), 703–705.

6 Endemic Aquatic Oligochaeta of India

6.1 INTRODUCTION

Aquatic endemic Oligochaeta include earthworms that are restricted to specific aquatic environments and survive in a freshwater or marine habitat. These species of earthworms are adapted to aquatic environments. Some species are endemic to specific regions and often relate to specific ecological conditions or the isolation of habitats.

Their habitat reveals preferences, including lakes, rivers, ponds, wetlands, and coastal zones. Many species have been recorded from specialized environments of caves or underground waters. They find importance from the ecological and taxonomic point of view due to their relatively high diversity and adaptability to different aquatic niches.

6.2 ENDEMIC SPECIES OF AQUATIC OLIGOCHAETES

Aquatic oligochaetes include two broad categories based on their nature of aquatic habitat:

1. Freshwater Oligochaeta
   They reside in the freshwater habitats of lakes, rivers, and ponds. A few are reported as endemic to specific water bodies or drainage basins. These include members of the following families:
   a. Family Naididae
      Family includes some of the most common freshwater species, including *Nais* spp. These species can often be highly endemic to specific regions or even certain lakes or rivers.
   b. Family Lumbriculidae
      This family includes species like *Lumbriculus* spp. It is commonly found in soft-bottom habitats in freshwater lakes and streams. Some species are endemic to specific river systems or freshwater regions.
   c. Family Deroidae
      They are found in fresh and brackish water habitats, and certain species are restricted to specific ecosystems or localities.
2. Marine Oligochaeta
   These species are restricted to the habitats of coastal waters, tidal zones, or even deep-sea environments. Some marine oligochaetes are highly endemic to specific coastal ecosystems or marine regions. These include members of:
   a. Family Orbiniidae: This family includes marine species residing in habitats of benthic environments, and some species are restricted to certain coastal areas.

DOI: 10.1201/9781003633600-6

b. Family Eunicidae: Marine oligochaetes of this family include some species that are endemic to particular oceanic regions.
c. Family Troglochaetidae: Some species of this family are restricted to caves or subterranean aquatic habitats, making them endemic to those regions.

Understanding the global scenario, Lake Baikal, a freshwater ecosystem in Russia, is reported to harbor endemic species of earthworm that have evolved in isolation over millions of years. They are important for researchers studying the mechanisms of speciation and adaptation to extreme conditions.

Some species of aquatic oligochaetes are found only in specific freshwater systems like the North American Great Lakes, which exhibit narrow distribution ranges, often confined to particular basins or tributaries, and they may be highly specialized for those environments.

Some Mediterranean freshwater systems host species of oligochaetes unique and endemic to those waters, including *Tubificoides mediterraneus*. Some species like *Troglochaetus beraneus* have been reported to be endemic to subterranean aquatic zones, residing in caves or underground water systems and dark environments. These species often exhibit adaptations like loss of pigmentation and reduced sensory structures.

India is rich and diverse in its flora and fauna. The diversity is also observed in the aquatic Oligochaeta species (Table 6.1), many of which are endemic to the country's diversely varied aquatic ecosystems. Earthworms are also known to play a critical role in the functioning of freshwater habitats, including rivers, lakes, wetlands, and marshes.

The rich biodiversity of India's aquatic ecosystems, from the Western Ghats to the northeastern wetlands, provides ideal conditions for the survival of these endemic species, many of which are adapted to specific environmental niches. Given the crucial role that aquatic oligochaetes play in maintaining ecosystem health, their conservation is of utmost importance. They are known for their contribution to essential ecological processes, including enhancement of soil fertility, purification of water, and playing a role in increasing the overall biodiversity.

Aquatic oligochaetes are integral to biotic and abiotic interactions within these ecosystems, contributing significantly to nutrient recycling, organic matter decomposition, and sediment bioturbation. They are known to play an important role in sustainable agriculture and soil health management by enhancing nutrient recycling, improving soil aeration, and promoting the decomposition of organic matter.

Their burrowing behavior helps improve soil and sediment structure by enhancing water infiltration and root growth and oxygenating the sediment, thus playing a major role in contributing to agricultural productivity. This in turn facilitates development of the plant root growth and supports microbial activity. They help in nutrient recycling by breaking down complex organic material into simpler compounds. This action also helps in enhancing the alteration of the substrate, which supports the growth of aquatic plants and the biodiversity of these habitats.

TABLE 6.1
List of Endemic Aquatic Oligochaeta of India

Order	Family	Species
Tubificida	Enchytraeidae	*Achaeta christenseni* Prabhoo, 1966
		Achaeta indica Prabhoo, 1960
		Achaeta nielseni Prabhoo, 1960
		Achaeta nurmineni Dash & Thambi, 1978
		Achaeta segmentata Prabhoo, 1966
		Enchytraeus albidus Henle, 1837
		Enchytraeus berhampurosus Dash & Thambi, 1978
		Enchytraeus indicus Stephenson, 1912
		Fridericia bulbosus Rosa, 1887
		Fridericia carmichaeli Stephenson, 1915
		Fridericia kalinga Dash et al., 1978
		Hemienchytraeus khallikotosus Dash & Thambi, 1978
		Hemienchytraeus theae Prabhoo, 1960
		Marionina indica Dash & Thambi, 1978
		Marionina macfadyeni Dash & Thambi, 1978
		Marionina subterranea (Knöllner, 1935)
		Stephensoniella marina (Moore, 1902)
	Naididae	*Aulophorus moghei* Naidu & Srivastava, 1980
		Nais andhrensis Naidu & Naidu, 1981

Research on the ecological role and distribution of endemic aquatic oligochaetes in India is a relatively new domain of study and has been developed in the last decade.

Studies as early as the 1990s by Julka and Paliwal (1993, 1994) emphasized the importance of these species in maintaining the health of freshwater ecosystems and their role in nutrient recycling.

Bhattacharya (2003) further explored the distribution patterns and ecological significance of these organisms, highlighting the uniqueness and endemic nature of many species found in different regions of India.

Sharma et al. (2020) identified new endemic aquatic Oligochaeta species from different river systems across the Indian subcontinent, revealing the remarkable adaptation of these species to India's diverse environmental conditions.

However, they face a significant threat from habitat destruction, pollution, climate change, and anthropogenic activities. The degradation of freshwater ecosystems due to agricultural runoff, industrial pollution, and urbanization poses a significant risk to these species, making it critical to implement effective conservation measures.

Studies by Julka (1994) and Sanyal (2007) have emphasized the need for habitat protection and sustainable land and water management practices to preserve these species.

The diverse climatic conditions and soil types in India support a variety of aquatic oligochaete species, and their conservation will be essential for the long-term sustainability of India's ecosystems.

However, more studies needs to be conducted on the distribution, ecological roles, and threats faced by these species that will aid in the development of effective conservation strategies, ensuring the health and vitality of India's freshwater ecosystems for future generations.

In this chapter, we highlight the current state of the art of endemic Oligochaeta of India, highlighting their distribution, ecological role, current status of conservation, and impact of anthropogenic interference and climate change on the endemic earthworm fauna of India.

6.3 DISTRIBUTION OF ENDEMIC OLIGOCHAETA IN INDIA

The distribution of endemic aquatic Oligochaeta species in India (Table 6.1, Figure 6.1) is highly concentrated in specific biodiversity hotspots, particularly in areas with distinct freshwater ecosystems.

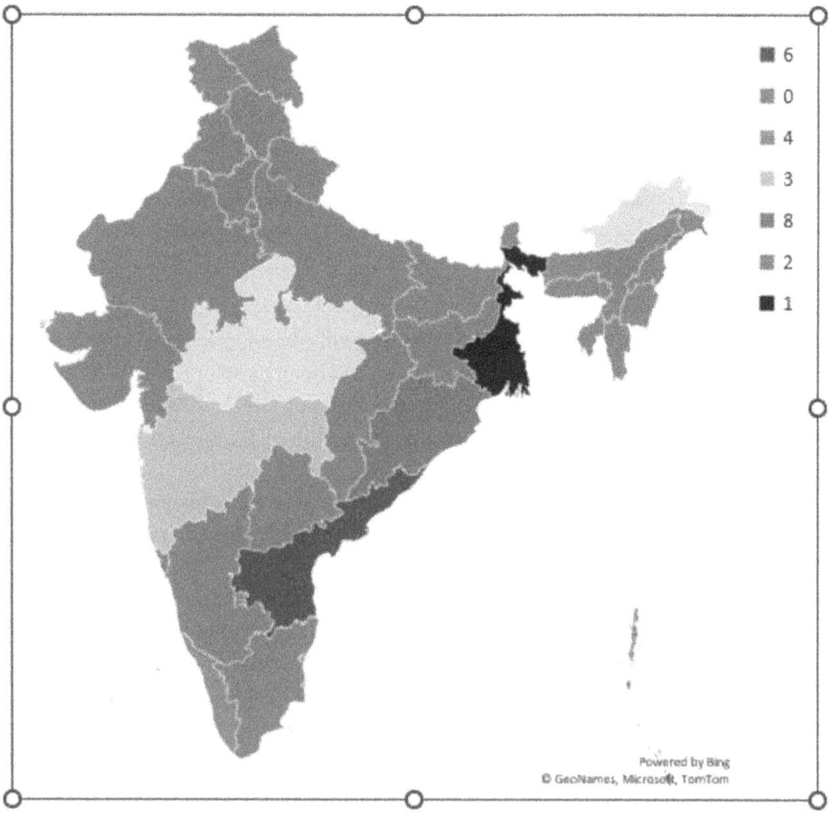

FIGURE 6.1 Distribution of aquatic endemic Oligochaeta in states of India: Andhra Pradesh, six species; Kerala, four species; Maharashtra, three species; Odisha, eight species; Tamil Nadu, two species; and West Bengal, one species.

(Balakrishnan et al., 2024).

In a recent study, indicated that several regions in India harbor a diverse range of endemic aquatic Oligochaeta species. While six endemic species have been recorded from the state of Andhra Pradesh, Kerala has been studied as the home to four species. Maharashtra, on the other hand, records three endemic species, and the state of Odisha stands out with eight endemic species. Tamil Nadu has two species, and West Bengal hosts one endemic species. These regions, characterized by their unique freshwater habitats, play a crucial role in supporting these specialized species. The findings highlight the importance of preserving these ecosystems for the conservation of endemic aquatic Oligochaeta diversity in India.

6.4 ECOLOGICAL ROLE OF ENDEMIC OLIGOCHAETA

Endemic aquatic Oligochaeta species play a crucial ecological role in maintaining the health and stability of freshwater ecosystems by contributing to soil aeration, organic matter decomposition, and nutrient cycling (Table 6.2). These species burrow into the sediment, facilitating the movement of water and air through the soil, which improves oxygenation and allows for better root growth in aquatic plants.

As detritivores, they break down organic matter, such as dead plant and animal material, into smaller particles, enriching the soil and sediment with essential nutrients. This decomposition process not only supports microbial communities but also enhances soil fertility, which in turn benefits other organisms within the ecosystem. In forests, wetlands, and agricultural lands, endemic aquatic Oligochaeta help recycle nutrients, improve the soil structure, and contribute to the overall productivity of the ecosystem.

Julka (1994) noted as early as 1994 that earthworms in wetlands help in the decomposition of plant material, which improves water quality by reducing organic load. Sanyal and Singh (2007), on similar lines of study, highlighted the role of these species in nutrient cycling and organic matter decomposition, showing their importance in maintaining ecosystem health and promoting biodiversity.

TABLE 6.2
Ecological Roles of Aquatic Endemic Oligochaetes

Decomposition	• **Play role in breaking down organic matter in aquatic ecosystems.** • **Feed on detritus, dead plant material, and microorganisms,** • **Help in recycling of nutrients back into the ecosystem.**
Sediment Bioturbation	• Many aquatic oligochaetes burrow in sediments, • This helps to aerate the soil, promote nutrient cycling, and improve the health of aquatic environments. • Finds importance in areas with low oxygen levels or poor water quality.
Food Source	• Form important food source for many aquatic organisms, including fish, crustaceans, and certain species of birds. • Play important role in maintaining the food web in aquatic ecosystems.

Their activities also help in reducing soil erosion and enhancing water retention. This is vital for the sustainability of agricultural practices and the health of wetlands.

6.5 THREATS TO ENDEMIC AQUATIC OLIGOCHAETA

Agricultural activities and activities from anthropogenic sources an climate changes have been studied as the major contributors to threats to endemic aquatic Oligochaeta (Table 6.3). We discuss the different threats in subsequent sections.

6.5.1 Agricultural Activities

Studies by Julka (1994) and Sanyal and Singh (2007) have highlighted the crucial role of these species in maintaining soil fertility and ensuring long-term agricultural sustainability. By promoting organic matter decomposition, aquatic earthworms also help reduce the need for synthetic fertilizers, contributing to more eco-friendly farming practices. Sustainable land and water management practices, such as reducing pesticide use, promoting organic farming, and conserving natural habitats, are essential to protect these valuable species and maintain the health of ecosystems that support agriculture (Julka, 1994; Sanyal & Singh, 2007).

But studies have revealed that the endemic aquatic Oligochaeta species are significantly impacted by agricultural practices, land use changes, and soil management strategies. Intensive agriculture, including large-scale use of chemical fertilizers, pesticides, and monoculture cropping, often leads to soil degradation, habitat destruction, and water pollution, directly affecting the survival of aquatic earthworms.

The runoff from agricultural fields containing these chemicals can contaminate freshwater habitats, reducing the populations of endemic Oligochaeta and other aquatic organisms. Changes in the usage of land like conversion of wetlands and forested areas into agricultural lands disrupts the natural ecosystems that support these species, leading to a decline in their populations.

TABLE 6.3
Potential Threats to Endemic Oligochaeta

Habitat destruction	**Pollution, damming, and land-use changes lead to loss of habitat.**
Agricultural activities	Use of pesticides; other toxic chemicals could leach into the water, leading to loss of their habitat.
Anthropogenic activities	Industrialization, altered land use, and release of pollutants into water could damage their habitat.
Invasive species	Introduction of non-native species into these ecosystems can outcompete or prey on endemic oligochaetes, causing the native population to decline.
Climate change	Changes in temperature, humidity, rainfall patterns, and water levels can affect the delicate balance of endemic species and their ecosystems.

Poor soil management practices, such as overgrazing or improper irrigation, also affect the structure and fertility of the soil and sediment, which in turn impacts earthworm habitats.

6.5.2 Anthropogenic Activities

Endemic aquatic Oligochaeta in India face several anthropogenic threats impacting their survival and distribution. Deforestation, particularly in wetland and riparian ecosystems, has led to habitat loss, altered hydrological conditions, disrupting the aquatic environments thereby impacting negatively the life and survival of these species (Srinivasan, 2006).

Industrialization, including pollution from untreated sewage and industrial effluents, has further degraded water quality, affecting the oxygen levels and chemical composition of aquatic habitats, which can be harmful to sensitive Oligochaeta populations (Jain et al., 2012).

The widespread use of pesticides in agriculture has contributed to chemical contamination in aquatic ecosystems, directly affecting the reproductive and survival rates of aquatic earthworms (Saxena & Verma, 2017).

Invasive species, such as non-native earthworm species introduced through horticultural activities, pose a significant threat to native Oligochaeta. These invasive species compete for resources, alter soil structure, and disrupt the ecological balance, further challenging the survival of endemic species (Sharma et al., 2015).

Such anthropogenic factors, along with climate change, continue to threaten the delicate balance of ecosystems where endemic aquatic Oligochaeta species thrive, emphasizing the need for conservation efforts to mitigate these impacts.

6.5.3 Climate Change

Climate change poses a significant threat by altering the hydrological patterns, temperature, and humidity levels, which can disrupt the delicate balance of aquatic ecosystems that support endemic Oligochaeta. We discuss the impact of climate change on both terrestrial and aquatic earthworms in this section.

6.6 CURRENT STATUS OF CONSERVATION

The conservation status of endemic aquatic Oligochaeta species in India is a growing concern due to several threats, including habitat destruction, pollution, and climate change. Many of these species are found in sensitive freshwater ecosystems, such as wetlands, rivers, and lakes, which are increasingly under threat from human activities.

Habitat destruction due to deforestation, land reclamation, and dam construction disrupts the natural habitats of aquatic earthworms, leading to a decline in their populations. Pollution, particularly from agricultural runoff containing pesticides and fertilizers, also negatively impacts the quality of water and soil, affecting the survival of these species.

Conservation efforts in India have been focused on protecting the habitats of these species through the establishment of protected areas and wetlands management.

Several biodiversity hotspots like the Western Ghats and northeastern wetlands have received attention for conservation initiatives.

Studies on the distribution and ecology of endemic aquatic earthworms, such as those conducted by Julka (1994) and Sanyal (2007), have contributed to a better understanding of their habitat needs and have informed conservation strategies.

Significant efforts to reduce pollution through better waste management and agricultural practices are being implemented in some regions to mitigate the impacts on freshwater habitats.

However, further research and stronger conservation measures are needed to ensure the long-term survival of endemic aquatic Oligochaeta species, as their role in ecosystem health is critical.

6.7 CONSERVATION AND MANAGEMENT STRATEGIES

The conservation of endemic aquatic Oligochaeta species requires a combination of in-situ and ex-situ strategies to ensure their survival and preserve their ecological roles. In-situ conservation measures focus on protecting and maintaining the natural habitats of these species.

This includes the establishment and management of protected areas, such as wetlands, rivers, and lakes, which serve as refuges for endemic aquatic earthworms. Conservation efforts should also prioritize the restoration of degraded aquatic ecosystems, particularly those impacted by pollution and habitat destruction. Implementing sustainable land and water management practices, such as reducing the use of harmful chemicals in agriculture, controlling deforestation, and promoting wetland conservation, is crucial to maintaining the health of freshwater habitats (Julka, 1994; Sanyal & Singh, 2007).

Ex-situ conservation strategies, such as the establishment of breeding programs in controlled environments, can also play an essential role in preserving these species. Aquatic Oligochaeta can be bred in laboratory conditions or aquatic reserves, where their populations can be monitored and managed to ensure genetic diversity and health (Sanyal, 2007).

Also, research initiatives are crucial for thoughtful the ecological requirements, distribution patterns, and threats to these species. Studies on the impact of climate change, pollution, and land-use changes on aquatic earthworms are necessary to develop effective management plans.

Collaboration between researchers, government agencies, and local communities can enhance the implementation of conservation strategies, and long-term monitoring programs will help assess the success of these measures (Julka, 1994; Sanyal, 2007). These efforts will help create a balanced approach to conservation that integrates both ecological protection and sustainable use.

6.8 DISCUSSION

Aquatic endemic Oligochaeta contribute significantly to the ecological integrity of their environments. Their specialization of habitat makes them intriguing for studies on biodiversity, speciation, and adaptation. Globally, they face challenges from environmental degradation and climate change, which can alter or even eliminate their unique habitats, thereby impacting their life and growth negatively. Conservation and management efforts focusing on protecting aquatic habitats find importance in maintaining the biodiversity of these ecosystems that are crucial for the continued survival of endemic Oligochaeta species.

The endemic aquatic Oligochaeta species of India represent a vital component of the country's rich freshwater biodiversity, playing essential roles in nutrient cycling, sediment bioturbation, and maintaining the overall health of aquatic ecosystems. Their presence in unique habitats, such as the Western Ghats, Eastern Himalayas, and northeastern wetlands, underscores the ecological significance of these regions and their role in fostering specialized species. However, these species face mounting threats from human activities, including habitat destruction, pollution, climate change, and invasive species.

As highlighted in the studies by Julka (1994), Sanyal (2007), and others, these anthropogenic pressures necessitate immediate and coordinated conservation efforts. Protecting and restoring freshwater habitats, implementing sustainable agricultural practices, and enhancing in-situ and ex-situ conservation measures are critical to safeguarding these species and the ecosystems they support.

Furthermore, continued research is essential to understand the distribution, ecology, and threats to aquatic Oligochaeta, which will inform effective conservation strategies. By prioritizing the protection of these endemic species, India can ensure the long-term health and sustainability of its freshwater ecosystems, which provide invaluable ecosystem services, such as water purification, soil fertility, and biodiversity conservation.

REFERENCES

Balakrishnan, S., Mondal, T., Mandal, C. K., Ahmed, S., Kumar, Y., Sivaleela, J. S. G., & Raghunathan, C. (2024). *Fauna of India Checklist, Annelida: Polychaete: Sipuncula: Clitella: Hirudinida: Crassiclitellata: Moniligastrida*. Online Version 1.0. Zoological Survey of India. https://doi.org/10.26515/Fauna/1/2023/Annelida: Polychaeta, Sipuncula, Clitella, Hirudinida, Crassiclitellata, Moniligastrida

Bhattacharya, S. (2003). Ecology and distribution of freshwater oligochaetes. *Journal of Aquatic Biology*, 18(2), 50–60.

Henle, F. G. J. (1837). Ueber Enchytraeus, eine neue Anneliden-Gattung. *Archiv für Anatomie, Physiologie und Wissenschaftliche Medicin*, 1837, 74–90.

Jain, P., Kumar, R., & Verma, S. (2012). Effects of industrial pollution on freshwater oligochaetes in India. *Water Quality and Ecology*, 45(1), 79–85.

Julka, J. M. (1994). Ecological role of earthworms in Indian wetlands. *Indian Journal of Ecology*, 21(3), 233–240.

Julka, J. M., & Paliwal, R. (1993). Freshwater oligochaetes of India: Their ecological importance and distribution. *Proceedings of the Indian National Science Academy*, 59(4), 387–397.

Julka, J. M., & R. Paliwal (1994). On a new species of Plutellus Perrier (Acanthodrilidae: Oligochaeta) from northwest Himalaya, India. *Research Bulletinof the Panjab University (Science)*, 44(I–IV), 217–220.

Knöllner, F. H. (1935). Ökologische und systematische Untersuchungen über litorale und marine Oligochäten der Kieler Bucht. *Zoologische Jahrbuecher (Systematik),* 66, 425–512.

Sanyal, A. (2007). Conservation and management of endemic freshwater earthworms in India. *Environmental Science and Technology*, 41(12), 4967–4973.

Sanyal, A., & Singh, A. (2007). The role of aquatic earthworms in nutrient cycling and ecosystem functioning. *Indian Journal of Soil Biology*, 38(4), 405–412.

Saxena, R., & Verma, M. (2017). Pesticide pollution in Indian freshwater habitats: Effects on aquatic earthworm populations. *Environmental Pollution*, 225, 230–238.

Sharma, P., Rathi, D., & Dutta, P. (2015). Invasive species and their impact on native Oligochaeta populations in India. *Journal of Aquatic Conservation*, 25(1), 82–90.

Sharma, V., Kaur, P., & Dey, S. (2020). Newly described endemic aquatic Oligochaeta species from the river systems of India. *International Journal of Aquatic Biology*, 11(5), 420–429.

Srinivasan, M. (2006). Impact of deforestation on freshwater ecosystems in India. *Environmental Conservation*, 33(2), 123–130.

7 Exotic Earthworm Species in India

7.1 INTRODUCTION

Exotic earthworm species in India have been a subject of significant ecological and agricultural interest, with both positive and negative implications. On the beneficial side, species like *Eisenia fetida* (commonly known as the red wiggler) and *Eudrilus eugeniae* (the African nightcrawler) have been widely utilized in vermiculture and organic farming practices. These species enhance soil fertility by accelerating the decomposition of organic matter, improving soil structure, and increasing nutrient availability, which is particularly valuable for sustainable agriculture (Bhattacharjee & Chaudhuri, 2002; Edwards, 2004).

Their role in waste management and soil enrichment has made them integral to eco-friendly farming systems in India. However, the introduction of exotic earthworms like *Pontoscolex corethrurus* has raised ecological concerns. This species, native to South America, has been reported to alter soil composition and disrupt native biodiversity in regions where it has been introduced. *P. corethrurus* is known for its ability to thrive in disturbed habitats, often outcompeting native earthworm species and modifying soil ecosystems by accelerating organic matter decomposition and altering nutrient cycling processes (Lavelle et al., 1987; Chaudhuri & Bhattacharjee, 2012). Such changes can negatively impact native flora and fauna, particularly in forest ecosystems, where soil dynamics play a critical role in maintaining ecological balance. The dual impact of exotic earthworms in India underscores the need for careful assessment and management of non-native species. While some species contribute significantly to agricultural productivity and waste management, others pose a threat to native ecosystems. This highlights the importance of species-specific studies and regulatory measures to mitigate potential ecological disruptions (Bhadauria & Ramakrishnan, 2007).

7.2 LIST OF 57 EXOTIC EARTHWORM SPECIES IN INDIA

Family: Lumbricidae

1. *Eisenia fetida* (Savigny, 1826)
2. *Lumbricus rubellus* Hoffmeister, 1843
3. *Lumbricus terrestris* Linnaeus, 1758
4. *Aporrectodea caliginosa* (Savigny, 1826)
5. *Aporrectodea trapezoides* (Dugès, 1828)
6. *Aporrectodea rosea* (Savigny, 1826)
7. *Dendrobaena veneta* (Rosa, 1886)
8. *Dendrodrilus rubidus* (Savigny, 1826)
9. *Octolasion cyaneum* (Savigny, 1826)
10. *Octolasion tyrtaeum* (Savigny, 1826)

DOI: 10.1201/9781003633600-7

Family: Megascolecidae

11. *Amynthas corticis* (Kinberg, 1867)
12. *Amynthas gracilis* (Kinberg, 1867)
13. *Metaphire houlleti* (Perrier, 1872)
14. *Metaphire posthuma* (Vaillant, 1868)
15. *Metaphire californica* (Kinberg, 1867)
16. *Perionyx excavatus* (Perrier, 1872)
17. *Perionyx sansibaricus* (Michaelsen, 1891a,b)
18. *Perionyx hawayana* (Rosa, 1891)
19. *Polypheretima elongata* (Perrier, 1872)
20. *Archipheretima montana* (Perrier, 1872)

Family: Eudrilidae

21. *Eudrilus eugeniae* (Kinberg, 1867)

Family: Glossoscolecidae

22. *Pontoscolex corethrurus* (Müller, 1857)
23. *Glossoscolex paulistus* (Michaelsen, 1918)

Family: Ocnerodrilidae

24. *Ocnerodrilus occidentalis* Eisen, 1878
25. *Eukerria saltensis* (Beddard, 1895a,b)
26. *Eukerria kuekenthali* (Michaelsen, 1908)

Family: Acanthodrilidae

27. *Dichogaster bolaui* (Michaelsen, 1891a,b)
28. *Dichogaster saliens* (Beddard, 1893)
29. *Dichogaster affinis* (Michaelsen, 1890)
30. *Dichogaster modiglianii* (Rosa, 1891)
31. *Dichogaster curgensis* (Gates, 1937)

Family: Octochaetidae

32. *Octochaetona beatrix* (Beddard, 1902)
33. *Octochaetona thurstoni* (Beddard, 1900)
34. *Octochaetona surya* (Gates, 1937)
35. *Octochaetona phillotti* (Beddard, 1901)

Family: Moniligastridae

36. *Drawida nepalensis* (Michaelsen, 1907)
37. *Drawida japonica* (Michaelsen, 1892)
38. *Drawida assamensis* (Stephenson, 1915)

Other Introduced Species

39. *Polypheretima elongata* (Perrier, 1872)
40. *Lampito mauritii* (Kinberg, 1867)
41. *Amynthas robustus* (Perrier, 1872)
42. *Eisenia andrei* (Bouché, 1972)
43. *Amynthas tokioensis* (Beddard, 1892)
44. *Perionyx macintoshi* (Stephenson, 1915)
45. *Eukerria sinensis* (Gates, 1930)
46. *Bimastos parvus* (Eisen, 1874)
47. *Amynthas diffringens* (Beddard, 1893)
48. *Dichogaster annae* (Beddard, 1896)
49. *Amynthas minimus* (Michaelsen, 1910)
50. *Dichogaster vulcani* (Horst, 1893)
51. *Metaphire birmanica* (Michaelsen, 1900)
52. *Metaphire bogorensis* (Michaelsen, 1892)
53. *Metaphire formosae* (Michaelsen, 1922)
54. *Pontoscolex latastii* (Beddard, 1894)
55. *Pontoscolex hebridensis* (Beddard, 1893)
56. *Amynthas hawayanus* (Rosa, 1891)
57. *Lampito paulistus* (Michaelsen, 1918)

7.3 IMPACT OF EXOTIC EARTHWORMS ON ECOSYSTEMS

The exotic earthworm species in India have led to a mix of beneficial and detrimental impacts on ecosystems and agriculture. Species such as *Eisenia fetida* and *Eudrilus eugeniae* have been widely recognized for their positive contributions to vermiculture and soil enrichment, enhancing organic matter decomposition and improving soil fertility, as highlighted by Sinha et al. (2010). These species have significantly boosted the vermicomposting industry, providing sustainable waste management solutions. However, exotic species like *Pontoscolex corethrurus* have raised ecological concerns, as they compete with native earthworm species, leading to their decline, and alter soil microhabitats, which negatively affects microbial communities, as reported by Chaudhuri & Bhattacharjee (2011). These disruptions to native biodiversity and soil composition underscore the need for careful management of exotic earthworm introductions to balance their agricultural benefits with ecological preservation.

7.4 DISCUSSION

Exotic earthworm species in India have brought about a complex interplay of benefits and challenges for both agriculture and natural ecosystems. On the one hand, species like *Eisenia fetida* and *Eudrilus eugeniae* have significantly contributed to sustainable farming practices by enhancing soil fertility, improving organic matter decomposition, and supporting the vermicomposting industry. These earthworms have become vital tools in organic waste management and soil enrichment, promoting eco-friendly agricultural systems. On the other hand, the invasion of species such as *Pontoscolex corethrurus* has raised serious ecological concerns. These exotic earthworms often outcompete native species, leading to their decline, and disrupt soil microhabitats, which can negatively impact microbial communities and alter nutrient cycling processes. Such changes pose a threat to native biodiversity, particularly in forest ecosystems where soil dynamics are critical for maintaining ecological balance. The dual impact of exotic earthworms highlights the need for careful assessment and management of their introduction. While their agricultural benefits are undeniable, their potential to harm native ecosystems calls for species-specific studies, regulatory measures, and monitoring systems to mitigate risks. Moving forward, it is essential to strike a balance between leveraging the advantages of exotic earthworms for sustainable agriculture and preserving the integrity of native ecosystems. This requires a proactive approach to ensure that the integration of non-native species is both ecologically responsible and beneficial for long-term agricultural productivity.

REFERENCES

Bhadauria, T., & Ramakrishnan, P. S. (2007). Earthworm diversity and ecosystem services in Indian agroecosystems. *Applied Soil Ecology*, 35(2), 406–418.

Bhattacharjee, G., & Chaudhuri, P. S. (2002). Earthworm resources of India and their utilization in vermicomposting. *Journal of Environmental Biology*, 23(3), 231–240.

Beddard, F. E. (1892). On some new species of earthworms from various parts of the world. *Proceedings of the Zoological Society of London*, 1892(2), 153–172.

Beddard, F. E. (1893). Microdrilus saliens Beddard. *Proceedings of the Zoological Society of London*, 1892, 683.

Beddard, F. E. (1894). A contribution to the knowledge of the Oligochaeta of South America. *Proceedings of the Zoological Society of London*, 1894, 250–257.

Beddard, F. E. (1895b). Preliminary account of new species of earthworms belonging to the Hamburg Museum. *Proceedings of the General Meetings for Scientific Business of the Zoological Society of London*, 63(2), 210–239.

Beddard, F. E. (1895a). *A monograph of the order of Oligochaeta* (769 pp.). Oxford: Clarendon Press.

Beddard, F. E. (1896). On some new species of earthworms from the islands of the Indian Ocean. *Proceedings of the Zoological Society of London*, 1895(4), 825–834.

Beddard, F. E. (1900). On a collection of earthworms from India and Ceylon. *Proceedings of the Zoological Society of London*, 1900, 856–868.

Beddard, F. E. (1901). On a new species of earthworm (Octochaetona phillotti) from the Khasia Hills, Assam. *Proceedings of the Zoological Society of London*, 1901(2), 269–271.

Beddard, F. E. (1902). Some new species of earthworms from various parts of the world. *Proceedings of the Zoological Society of London*, 1902(1), 137–170.

Bouché, M. B. (1972). Lombriciens de France. Écologie et Systématique. *Annales de Zoologie-Écologie Animale*, (Numéro Hors-Série), 1–671.

Chaudhuri, P. S., & Bhattacharjee, G. (2011). Competence of exotic and native earthworm species in vermicomposting. *Environmentalist*, 31(4), 394–401.

Chaudhuri, P. S., & Bhattacharjee, G. (2012). Impact of exotic earthworms on soil microbial communities and native species in India. *Soil Biology and Biochemistry*, 45, 1–8.

Dugès, A. (1828). Recherches sur les vers intestins et quelques espèces d'helminthes. *Annales des Sciences Naturelles*, Paris, 15, 284–305.

Edwards, C. A. (2004). *Earthworm Ecology* (2nd ed.). CRC Press.

Eisen, G. (1874). On the Anatomy and Systematics of Oligochaeta: I. New American Lumbricidae. *Öfversigt af Kongliga Vetenskaps-Akademiens Förhandlingar*, 31(2), 51–97.
Gates, G. E. (1930). On some earthworms from Szechwan and Hupeh. *Journal of the Proceedings of the Asiatic Society of Bengal, New Series*, 26, 241–247.
Gates, G. E. (1937). On a collection of Indian earthworms. *Records of the Indian Museum*, 39(4), 405–497.
Horst, R. (1893). On a collection of Earthworms from Sumatra. *Notes from the Leyden Museum*, 15(3-4), 195–214.
Kinberg, J. G. H. (1867). Annulata nova. *Öfversigt af Kongliga Vetenskaps-Akademiens Förhandlingar*, 23(4), 97–103.
Lavelle, P., Bignell, D., & Lepage, M. (1987). Soil function in a changing world: The role of invertebrate ecosystem engineers. *European Journal of Soil Biology*, 33(4), 159–193.
Michaelsen, W. (1890). Arhropoden. In: *Die Forschungsreise S.M.S. "Gazelle"* (Eds. Möbius, K., & Hertwig, R.), vol. 2, pp. 1–71. Mittheilungen aus dem Naturhistorischen Museum in Hamburg.
Michaelsen, W. (1891a). Beschreibung der von Herrn Dr. Fr. Stuhlmann auf Sansibar und dem gegenüberliegenden Festlande gesammelten Terricolen. Anhang: I. Übersicht über die Teleudrilinen. II. Die Terricolen-Fauna Afrikas. *Jahrbuch der Hamburgischen Wissenschaftlichen Anstalten*, 9, 3–72.
Michaelsen, W. (1891b). Terricolen der Berliner Zoologischen Sammlungen. *Archiv für Naturgeschichte*, 57(1), 33–62.
Michaelsen, W. (1892). Terricolen der Berliner Zoologischen Sammlungen II. *Archiv für Naturgeschichte*, 58(1), 209–261.
Michaelsen, W. (1900). Oligochaeta. In W. F. H. Blandford (Ed.), *The Fauna of British India, including Ceylon and Burma*. Taylor and Francis, London.
Michaelsen, W. (1907). Neue Oligochäten von Vorder-Indien, Ceylon, Burma und den Andaman-Inseln. *Mitteilungen aus dem Naturhistorischen Museum in Hamburg*, 24, 143–193.
Michaelsen, W. (1908). Die Oligochäten Westindiens. *Zoologische Jahrbücher, Abteilung für Systematic, Jena, Supplement*, 11, 13–32, 1Taf.
Michaelsen, W. (1910). Die Oligochäten Fauna Neuguineas. *Abhandlungen herausgegeben vom Naturwissenschaftlichen Verein zu Hamburg*, 19(4), 1-32.
Michaelsen, W. (1918). Die Lumbriciden. *Zoologische Jahrbücher. Abteilung für Systematik, Geographie und Biologie der Tiere*, 41, 1–398.
Michaelsen, W. (1922). Neue Oligochäten von den Inseln der Sunda-Archipels. *Arkiv för Zoologi*, 14(8), 1–11.
Müller, F. (1857). Lumbricus corethrurus, Burstenschwanz. *Archiv für Naturg*, 23, 6–113.
Perrier, E. (1872). Recherches pour servir à l'histoire des Lombriciens terrestres. *Nouvelles Archives du Muséum d'Histoire Naturelle*, Paris, 8, 5–198.
Rosa, D. (1886). Allolobophora veneta, n. sp. (Lumbricidae). *Bollettino dei Musei di Zoologia ed Anatomia Comparata della Reale Università di Torino*, 1(11), 674.
Rosa, D. (1891). Die exotischen Terricolen des K. K. Naturhistorischen Hofmuseums in Wien. Annalen des K. K. *Naturhistorischen Hofmuseums in Wien*, 6, 379–406.
Savigny, J. C. (1826). [Annelida]. In Cuvier, G. (Ed.) *Analyse des travaux de l'Académie royale des Sciences pendant l'année 1821, partie physique* (vol. 5, pp. 176–184). Mémoires de l'Académie (royale) des sciences de l'Institut (impérial) de France.
Sinha, R. K., Valani, D., Chauhan, K., & Agarwal, S. (2010). Embarking on a second green revolution for sustainable agriculture by vermiculture biotechnology. *Journal of Agricultural and Environmental Ethics*, 23(4), 371–392.
Stephenson, J. (1915). On some Indian Oligochaeta, mainly from the collection of the Indian Museum. *Records of the Indian Museum*, 11(3), 295–353.
Vaillant, L. (1868). Note sur le *Perichaeta posthuma. Annales de la Société Entomologique de France*, 4(8), 137–140.

8 Semiaquatic Earthworm Fauna

8.1 INTRODUCTION

Megadrile earthworms are the majority of animal biomass in soil ecosystems (Fragoso et al., 1999). The earthworm fauna of India are currently represented by 459 species distributed across ten families: Moniligastridae, Lumbricidae, Almidae, Rhinodrilidae, Acanthodrilidae, Eudrilidae, Ocnerodrilidae, Benhamiidae, Octochaetidae, and Megascolecidae (Balakrishnan et al., 2024).

A unique feature of some semiaquatic earthworms, such as those in the genus *Glyphidrilus*, is the presence of peculiarly expanded epidermal structures near the clitellum, referred to as "wings". Although the function of these wings is not well understood, Gates (1972) proposed their probable evolution as vascularized organs for respiration in aquatic habitats. However, given that these structures are often located deep within burrows where oxygen levels are low, an alternative explanation suggests they may play a role in copulation, similar to the *Tubercula pubertatis* found in some other earthworm families. This highlights the need for further research to fully understand the ecological and physiological roles of these adaptations in semiaquatic earthworms.

Despite occupying only about 2% of the world's total land area, India is home to approximately 10.5% of the globally known earthworm species (Julka et al., 2009). The level of endemism is remarkably high, with around 71% of genera and 85% of earthworm species being native to India (Julka & Paliwal, 2005). Among the *Glyphidrilus* genus, 40 species-level taxa are found in China, Cambodia, India, Indonesia, Laos, Myanmar, Sri Lanka, Singapore, and Thailand, but only one species, *G. stuhlmanni* Michaelsen, is recorded from Africa (Chanabun et al., 2013) (Table 8.1, Figures 8.1–8.12).

TABLE 8.1
Semiaquatic Earthworm Species in India

Family	Genus	Species
Almidae	Glyphidrilus	*Glyphidrilus annandalei* Michaelsen, 1910
Almidae	Glyphidrilus	*Glyphidrilus elegans* Rao, 1922
Almidae	Glyphidrilus	*Glyphidrilus fluviatilis* Rao, 1922
Almidae	Glyphidrilus	*Glyphidrilus gangeticus* Gates, 1958
Almidae	Glyphidrilus	*Glyphidrilus papillatus* (Rosa, 1890)
Almidae	Glyphidrilus	*Glyphidrilus spelaeotes* Stephenson, 1924
Almidae	Glyphidrilus	*Glyphidrilus tuberosus* Stephenson, 1916

DOI: 10.1201/9781003633600-8

FIGURE 8.1 (a) General characteristics of *Glyphidrilus* Horst, 1889; (b) tail tips of two individuals of *Glyphidrilus* in the normal position on the soil surface; and (c) *Glyphidrilus* cocoons.

(Image adapted from Chanabun et al., 2013 under open access article distributed under the terms of the Creative Commons Attribution License 3.0 (CC-BY)).

FIGURE 8.2 Photographs of (a) *Glyphidrilus nanensis* sp. n. casts; (b) type locality of *G. nanensis* sp. n. in the rice field near Nan River at Saklek, Phichit, north Thailand; (c) coloration of living paratype.

(Image adapted from Chanabun et al., 2017 under open access article distributed under the terms of the Creative Commons Attribution License 3.0 (CC-BY)).

FIGURE 8.3 Images of (a) *Glyphidrilus chiangraiensis* sp. n. and other earthworms casts; (b) type locality of *G. chiangraiensis* sp. n. in the river banks of Mekong River at Wat Hatkai, Chiangkhong, Chiangrai, north Thailand, and (c) newly collected specimen after the first preservation step in 30% (v/v) ethanol.

(Image adapted from Chanabun et al., 2017 under open access article distributed under the terms of the Creative Commons Attribution License 3.0 (CC-BY)).

FIGURE 8.4 Images of (a) *Glyphidrilus champasakensis* sp. n.; (b) type locality of *G. champasakensis* sp. n. in the banks of Mekong River at Ban Khonkhen, Champasak, Laos; and (c) newly collected specimen after the first preservation step in 30% (v/v) ethanol.

(Image adapted from Chanabun et al., 2017 under open access article distributed under the terms of the Creative Commons Attribution License 3.0 (CC-BY)).

FIGURE 8.5 Images of (a) *Metaphire* sp., (b) *Amynthas mekongianus* (Cognetti, 1922), (c and d) cocoons of *Glyphidrilus* in the soil.

(Image adapted from Chanabun et al., 2017 under open access article distributed under the terms of the Creative Commons Attribution License 3.0 (CC-BY)).

FIGURE 8.6 Images of (a and b) type locality of *Glyphidrilus borealis* sp. n.(arrow head), a muddy swamp near Maeklang Waterfall, Doi Inthanon National Park, Chiangmai, and (c) coloration of newly collected paratype CUMZ 3224 just after the first preservation step in 30% (v/v) ethanol.

(Image adapted from Chanabun et al., 2013 under open access article distributed under the terms of the Creative Commons Attribution License 3.0 (CC-BY)).

FIGURE 8.7 Images of (a) type locality of *Glyphidrilus vangthongensis* sp. n.(arrow head) on a river bank of Sakulnothayan waterfall, Vangthong, Phitsanulok; (b) *Glyphidrilus vangthongensis* sp. n. casts, and (c) newly collected just after the first preservation step in 30% (v/v) ethanol.

(Image adapted from Chanabun et al., 2013 under open access article distributed under the terms of the Creative Commons Attribution License 3.0 (CC-BY)).

FIGURE 8.8 Images of (a) type locality of *Glyphidrilus chaophraya* sp. n. on a river bank of Chaophraya River, Payuhakiri, Nakhonsawan; (b) tail tip; and (c) newly collected after the first preservation step in 30% (v/v) ethanol.

(Image adapted from Chanabun et al., 2013 under open access article distributed under the terms of the Creative Commons Attribution License 3.0 (CC-BY)).

FIGURE 8.9 Images of (a) type locality of *Glyphidrilus quadratus* sp. n. on a sandy beach at Kang Saphue, Phibunmangsahan, Ubon Ratchathani; (b) *Glyphidrilus quadratus* sp. n. casts; and (c) coloration of newly collected after the first preservation step in 30% (v/v) ethanol.

(Image adapted from Chanabun et al., 2013 under open access article distributed under the terms of the Creative Commons Attribution License 3.0 (CC-BY)).

FIGURE 8.10 Images of (a) type locality of *Glyphidrilus wararamensis* sp. n. in a muddy swamp at a stream near Wat Tham Wararam, Phanom, Suratthani; (b) *Glyphidrilus wararamensis* sp. n. casts; and (c) newly collected just after the first preservation step in 30% (v/v) ethanol.

(Image adapted from Chanabun et al., 2013 under open access article distributed under the terms of the Creative Commons Attribution License 3.0 (CC-BY)).

FIGURE 8.11 Images of (a and b) type locality of *Glyphidrilus vesper* sp. n. in a muddy swamp near the Thilorsu Waterfall, Umphang, Tak, and (c) coloration of newly collected paratype CUMZ 3267 just after the first preservation step in 30% (v/v) ethanol.

(Image adapted from Chanabun et al., 2013 under open access article distributed under the terms of the Creative Commons Attribution License 3.0 (CC-BY)).

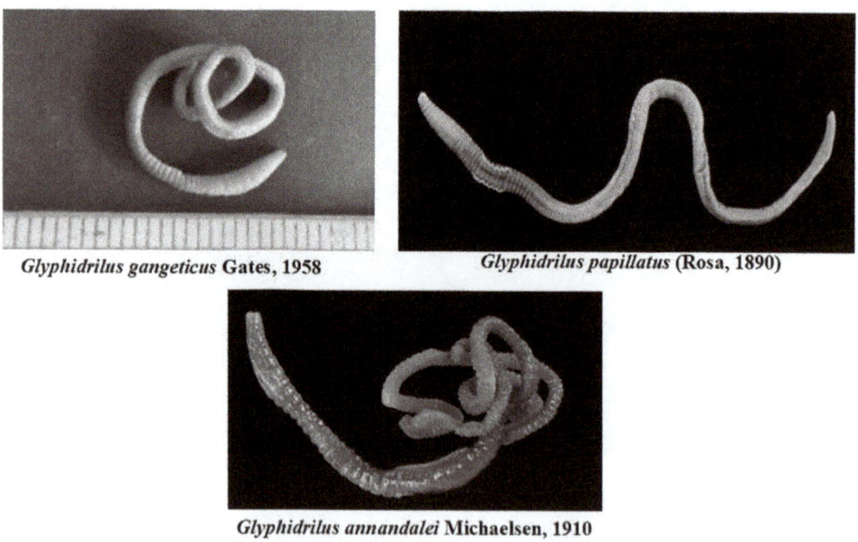

FIGURE 8.12 Semi-aquatic earthworms, *Glyphidrilus gangeticus* Gates, 1958, *Glyphidrilus papillatus* (Rosa, 1890), and *Glyphidrilus annandalei* (Michaelsen, 1910), from India.

(Image reproduced with permission of director, Zoological Survey of India).

8.2 HABITATS OF SEMIAQUATIC EARTHWORMS

Semiaquatic earthworm species are predominantly found in waterlogged or moist habitats such as riverbanks, wetlands, paddy fields, and marshy areas. These environments provide the ideal conditions for their survival and proliferation, as they require consistent moisture and organic-rich substrates. Semiaquatic earthworms play a vital ecological role in these habitats by enhancing organic matter decomposition, which facilitates nutrient recycling and improves soil fertility.

Additionally, their burrowing activities promote soil aeration, aiding root growth and microbial activity. Their unique adaptation to semiaquatic environments enables them to thrive in areas with fluctuating water levels, such as floodplains and seasonal wetlands, where they contribute significantly to maintaining soil structure and health.

These earthworms are particularly important in agricultural systems like paddy fields, where their activities support sustainable farming practices by improving soil quality and nutrient availability. The ecological functions of these species underscore their importance in maintaining the balance of semiaquatic ecosystems and their contribution to soil health and productivity (Gates, 1958; Rao, 1922; Stephenson, 1916).

8.3 ECOLOGICAL ROLE OF SEMIAQUATIC EARTHWORM

Semiaquatic earthworms, under the *Glyphidrilus* genus, have crucial ecological importance due to their role in waterlogged moist habitats, including marshy areas, riverbanks, wetlands, and paddy fields. These earthworms significantly enhance organic matter decomposition, facilitating nutrient cycling and improving soil fertility. Their burrowing activities promote soil aeration, which benefits root growth and microbial activity, thereby enhancing overall soil health. In agricultural systems like paddy fields, semiaquatic earthworms contribute to sustainable farming by improving soil structure and nutrient availability. Their adaptation to fluctuating water levels allows them to thrive in dynamic environments, where they help maintain ecological balance and support biodiversity. These functions highlight their importance in nutrient cycling, soil enrichment, and ecosystem stability in semiaquatic habitats (Gates, 1958; Rao, 1922; Stephenson, 1916).

8.4 SEMIAQUATIC EARTHWORMS

The study of semiaquatic earthworms in India, particularly those belonging to the genus *Glyphidrilus*, highlights their ecological significance and unique adaptations to waterlogged and moist habitats. Until 2013, 25 species and 1 subspecies of *Glyphidrilus* were recorded (Chanabun et al., 2013), all from Asia and Africa. However ongoing research is adding new records of semiaquatic earthworms under this genus. These earthworms are integral to the functioning of ecosystems such as riverbanks, wetlands, paddy fields, and marshy areas.

In 2017, globally, seven new species of semi-aquatic freshwater earthworms under genus *Glyphidrilus* Horst, 1889 were reported from Thailand and Laos (Chanabun et al., 2017), including *Glyphidrilus nanensis* Chanabun & Panha sp. n., *G. satunensis* Chanabun & Panha sp. n., *G. chiangraiensis* Chanabun & Panha sp. n., *G. namphao* Chanabun & Panha sp. n., *G. sekongensis* Chanabun & Panha sp. n., *G. namdonensis* Chanabun & Panha sp. n., and *G. champasakensis* Chanabun & Panha sp. n. Searches for new records and species are enriching the total list of semiaquatic earthworms globally. We present pictorial images of some species under genus *Glyphidrilus* for readers.

8.5 DISCUSSION

Semiaquatic earthworms are known to play a vital role in the decomposition of organic matter, nutrient cycling, and aeration of soil, which are critical for maintaining soil fertility and structure. Their burrowing activities enhance microbial activity and root growth, contributing to the overall health of the ecosystems they inhabit. In agricultural systems, particularly paddy fields, they support sustainable farming practices by improving soil quality and nutrient availability. Their ability to thrive in environments with fluctuating water levels underscores their ecological versatility and importance.

Despite their ecological benefits, the unique adaptations of these earthworms, such as the vascularized "wings" near the clitellum, remain poorly understood, warranting further research to elucidate their functional roles. The high endemism of earthworm species in India, coupled with their ecological contributions, underscores the need for conservation efforts to protect these vital organisms and their habitats.

As India continues to face challenges related to soil degradation and sustainable agriculture, semiaquatic earthworms offer a natural solution for enhancing soil health and ecosystem stability. Future studies should focus on understanding their ecological interactions, habitat requirements, and potential responses to environmental changes to ensure their conservation and continued contribution to ecosystem services. This research not only enriches our understanding of semiaquatic earthworm biodiversity but also emphasizes their role in maintaining ecological balance and supporting sustainable agricultural practices in India.

Members of genus *Glyphidrilus* from Southeast Asia have been studied to reveal morphological crypsis in divergent phylogenetic lineages and a high morphological variability in the same phylogenetic lineages. Recently, therefore, attempts are being made to study the complex genus by integrating DNA sequence and morphological data (Jirapatrasilp et al., 2019). However, more research is required to understand the diversity and complexity of members of this genus.

REFERENCES

Balakrishnan, S., Mondal, T., Mandal, C. K., Ahmed, S., Kumar, Y., Sivaleela, J. S. G., & Raghunathan, C. (2024). *Fauna of India Checklist, Annelida: Polychaete: Sipuncula: Clitella: Hirudinida: Crassiclitellata: Moniligastrida.* Online Version 1.0. Zoological Survey of India. https://doi.org/10.26515/Fauna/1/2023/Annelida

Chanabun, R., Inkavilay, K., & Panha, S. (2017). New species of semi-aquatic freshwater earthworm genus *Glyphidrilus* Horst, 1889 from Thailand and Laos (Oligochaeta, Almidae). *ZooKeys*, 672, 1–34. https://doi.org/10.3897/zookeys.672.11809

Chanabun, R., Sutcharit, C., Tongkerd, P., & Panha, S. (2013). The semi-aquatic freshwater earthworms of the genus *Glyphidrilus* Horst, 1889 from Thailand (Oligochaeta, Almidae) with re-descriptions of several species. *ZooKeys*, 265, 1–76. https://doi.org/10.3897/zookeys.265.4454

Cognetti, L. (1922). Oligocheti dell'Annam. *Bollettino dei Musei di Zoologia ed Anatomia Comparata della Reale Università di Torino*, 37(748), 1–19.

Fragoso, C., Brown, G. G., Patrón, J. C., Blanchart, E., Lavelle, P., Pashanasi, B., & Senapati, B. K. (1999). Earthworm communities of tropical agroecosystems: Origin, structure, and influence of management practices. *Applied Soil Ecology*, 13(2), 137–153. https://doi.org/10.1016/S0929-1393(99)00028-6

Gates, G. E. (1958). On some earthworms from India. *Proceedings of the Indian Academy of Sciences*, 47(3), 177–189.

Gates, G. E. (1972). Burmese earthworms: An introduction to the systematics and biology of megadrile oligochaetes with special reference to Southeast Asia. *Transactions of the American Philosophical Society*, 62(7), 1–326. https://doi.org/10.2307/1006216

Jirapatrasilp, P., Backeljau, T., Prasankok, P., Chanabun, R., & Panha, S. (2019). Untangling a mess of worms: Species delimitations reveal morphological crypsis and variability in Southeast Asian semi-aquatic earthworms (Almidae, *Glyphidrilus*). *Molecular Phylogenetics and Evolution*, 139, 106531. https://doi.org/10.1016/j.ympev.2019.106531

Julka, J. M., & Paliwal, R. (2005). Endemism and diversity of Indian earthworms. *Proceedings of the National Academy of Sciences, India*, 75(3), 211–220.

Julka, J. M., Paliwal, R., & Kathireswari, P. (2009). Earthworm resources of India and their utilization in vermiculture. *Indian Journal of Ecology*, 36(1), 1–10.

Rao, C. R. N. (1922). On some new species of earthworms from South India. *Records of the Indian Museum*, 24(3), 291–310.

Rosa, D. (1890). Monoligastridi, Geoscolecidi ed Eudrilidi. *Annali del Museo Civico di Storia Naturale di Genova*, 29(2), 365–398.

Stephenson, J. (1916). On some Indian Oligochaeta, mainly from the collection of the Indian Museum. *Records of the Indian Museum*, 12(3), 299–348.

Stephenson, J. (1924). *The Fauna of British India, Including Ceylon and Burma: Oligochaeta*. Taylor & Francis.

9 Morphology and Anatomy, Behavior

9.1 INTRODUCTION

Earthworms show distinct morphological and anatomical features that are vital for their ecological functions. Externally, their body is segmented, with each segment marked by ring-like annuli, providing flexibility and aiding in movement (Edwards & Bohlen, 1996). The clitellum, a thickened glandular region typically found between the 14th and 16th segments, plays an essential role in reproduction by secreting mucus to form cocoons (Julka, 1988). Setae, tiny bristle-like structures, are arranged in pairs on each segment and aid in anchoring the worm during movement and burrowing (Darwin, 1881). Internally, earthworms possess a well-developed digestive system comprising the buccal cavity, pharynx, esophagus, crop, gizzard, and intestine, specialized for decomposing organic matter (Howmiller & Beeton, 1971). Their closed circulatory system includes dorsal and ventral blood vessels, with aortic arches functioning as hearts (Reynolds & Wetzel, 2024). The nervous system consists of a cerebral ganglion (brain) and a ventral nerve cord, coordinating sensory and motor activities (Julka & Paliwal, 1993). Reproductively, earthworms are hermaphroditic, possessing both male and female reproductive organs, and mating involves mutual exchange of sperm (Stephenson, 1930). Locomotion and burrowing are facilitated by the coordinated contraction and relaxation of circular and longitudinal muscles, aided by setae anchoring segments in the soil (Edwards & Bohlen, 1996). This burrowing behavior enhances soil aeration and nutrient cycling (Darwin, 1881).

9.2 DIFFERENCES BETWEEN MORPHOLOGY AND ANATOMY

The morphology and anatomy of earthworms differ primarily in their focus on external versus internal structures, respectively. Morphology refers to the observable external features, such as body segmentation, the clitellum, and the arrangement of setae, which are decisive for movement, reproduction, and environmental interaction (Darwin, 1881; Edwards & Bohlen, 1996). Anatomy, on the other hand, pertains to the internal structures, including the digestive, circulatory, and nervous systems, which support the organism's physiological functions (Howmiller & Beeton, 1971; Julka & Paliwal, 1993). The internal anatomy is specialized for the digestion of organic matter and nutrient cycling, with a closed circulatory system and a ventral nerve cord (Reynolds & Wetzel, 2024). While morphology directly influences the earthworm's interaction with its environment, anatomy is responsible for the internal processes that sustain life, prominence the complementary roles. These two aspects play a role in the organism's survival and ecological contribution.

DOI: 10.1201/9781003633600-9

9.3 MORPHOLOGY OF ORGANISMS

The morphology of earthworms is characterized by several distinct external features that aid in their survival and ecological functions. Their body is segmented, with each segment separated by annuli, which provides flexibility and facilitates movement through soil (Edwards & Bohlen, 1996). The clitellum, a glandular structure located between the 14th and 16th segments, plays a crucial role in reproduction by secreting mucus to form a cocoon for fertilization and embryo development (Julka, 1988). Setae, small bristle-like structures arranged in pairs on each segment, anchor the earthworm during burrowing and locomotion, allowing it to move through the soil effectively (Darwin, 1881). This external morphology is well adapted for the earthworm's burrowing lifestyle, enabling it to efficiently move, reproduce, and interact with its environment, contributing to soil aeration and nutrient cycling.

9.4 ANATOMY OF ORGANISMS

The anatomy of the earthworm consists of specialized internal structures that enable it to perform essential physiological functions. The digestive system is well developed, comprising the buccal cavity, pharynx, esophagus, crop, gizzard, and intestine, all of which work together to decompose organic matter and absorb nutrients (Howmiller & Beeton, 1971). Earthworms possess a closed circulatory system, with dorsal and ventral blood vessels and a series of aortic arches that function as hearts, ensuring the circulation of nutrients and gases throughout the body (Reynolds & Wetzel, 2024). The nervous system includes a cerebral ganglion (brain) and a ventral nerve cord, which coordinates sensory input and motor output, allowing the earthworm to respond to environmental stimuli (Julka & Paliwal, 1993). Reproductively, earthworms are hermaphrodites, possessing both male and female reproductive organs, and engage in mutual sperm exchange during mating (Stephenson, 1930). These anatomical features support the earthworm's ecological role in soil health by facilitating its digestion, circulation, coordination, and reproduction.

9.5 COMPARATIVE MORPHOLOGY AND ANATOMY

The comparative morphology and anatomy of earthworms reveal how external and internal features work together to support their ecological role. Externally, the segmented body, marked by annuli, provides flexibility for movement through soil, while the clitellum aids in reproduction by secreting mucus for cocoon formation (Darwin, 1881; Julka, 1988). Internally, the digestive system is specialized for processing organic matter, and the closed circulatory system, with aortic arches functioning as hearts, efficiently transports nutrients (Howmiller & Beeton, 1971; Reynolds & Wetzel, 2024). The nervous system coordinates movement and environmental response via the cerebral ganglion and ventral nerve cord (Julka & Paliwal, 1993). A comparative analysis of these features highlights how the earthworm's external morphology is intricately linked to its internal anatomy, with structural adaptations facilitating burrowing, nutrient absorption, and reproduction, which in turn contribute to the soil's health and fertility.

9.5.1 Functional Significance

Earthworms enrich soil ecosystems, significantly influencing soil structure, fertility, and overall ecosystem health. From an anatomical and physiological perspective, their digestive system, particularly the gizzard and intestine, enables the efficient breakdown of organic matter, enhancing nutrient cycling. Earthworms also possess a specialized circulatory system with a closed circulatory loop, enabling the distribution of nutrients and oxygen throughout their body. In terms of physiology, earthworms display biochemical processes, such as the secretion of enzymes in the digestive tract that assist in breaking down complex organic compounds into simpler forms. According to Edwards and Bohlen (1996), earthworms are vital bioindicators of soil health, as their presence and abundance can reflect the environmental conditions and the quality of the habitat. Their burrowing behavior further contributes to soil aeration and water infiltration, improving soil structure and promoting plant growth.

9.6 BEHAVIOR

Earthworm behavior is shaped by various environmental factors, and understanding their actions is crucial to appreciating their ecological significance. Earthworms play an essential role in soil aeration, nutrient cycling, and organic matter decomposition. They create extensive burrows while moving through the soil, which improves soil structure and fertility by allowing air and water to permeate the soil. The depth and soil texture preferences of earthworms vary across species. For example, *Lumbricus terrestris* prefers deep, moist soils, while other species may stay closer to the surface (Parle, 1962). Soil moisture and temperature significantly affect burrowing depth, with earthworms moving deeper during dry periods to avoid desiccation (Lowe & Butt, 2014). Earthworms feed primarily on decaying plant matter, converting it into humus and contributing to nutrient recycling (Zhang & Xue, 2011). They are selective feeders, preferring organic material rich in nutrients and microbial activity, which helps optimize energy intake and reproductive success (Schulz & Zander, 2017). Earthworms are highly sensitive to moisture and temperature, avoiding desiccation by seeking out moist conditions (Bouché, 1977), and their activity is most prominent between 10°C and 20°C (Edwards, 2004). They also avoid light to protect themselves from predators, a behavior known as negative phototropism (Denny, 1980). Earthworms are hermaphroditic and mate in moist conditions, exchanging sperm for fertilization. After mating, they produce cocoons where fertilized eggs develop, with environmental factors like food availability influencing egg number (Hartenstein, 1982; Blouin et al., 2013). Earthworms tend to aggregate in areas with high organic matter or favorable conditions, which can lead to increased reproductive success (González et al., 2019), although they may compete for space and food resources in dense populations (Smith, 2015). To avoid predators like birds and amphibians, earthworms retreat into their burrows or secrete chemical defenses, such as slime or noxious substances (Gist & Gower, 1987). Additionally, earthworms exhibit a behavioral response to vibrations, retreating deeper into the soil when they sense potential threats (Schmidt & Zangerlé, 2001). Earthworms display high behavioral plasticity, adapting to environmental changes by migrating deeper during droughts or moving to the surface during flooding (Edwards & Bohlen, 1996). These behaviors demonstrate earthworms' critical role in maintaining soil health and their ability to adapt to environmental changes.

9.7 BURROWING BEHAVIOR

Earthworms create extensive burrows as they move through the soil, consuming organic material and contributing to soil aeration, which enhances soil structure and fertility (Edwards & Bohlen, 1996). These burrows also provide vital pathways for water and air to penetrate the soil. Earthworms display species-specific preferences for soil texture and depth; for instance, species like *Lumbricus terrestris* prefer deeper, moist soils, while others may remain near the surface under more temperate conditions (Parle, 1962). Burrowing depth is influenced by soil moisture and temperature, with earthworms retreating to deeper soil layers during dry or extreme conditions to maintain moisture and avoid desiccation (Lowe & Butt, 2014).

9.8 FEEDING BEHAVIOR

Earthworms play a crucial role in the decomposition of organic matter by feeding primarily on decaying plant material, including dead leaves and roots. This feeding behavior contributes significantly to the conversion of organic matter into nutrient-rich humus, thereby facilitating nutrient recycling in the soil (Zhang & Xue, 2011). Moreover, research indicates that earthworms are selective in their feeding habits, preferring organic matter with higher nutrient content and greater microbial activity. This selective feeding behavior helps them optimize energy efficiency, ensuring better growth, survival, and reproductive success (Schulz & Zander, 2017). By feeding on organic matter that is rich in nutrients and microbes, earthworms contribute to soil fertility while simultaneously benefiting from an energy-efficient feeding strategy that supports their reproductive and metabolic processes.

9.9 RESPONSE TO ENVIRONMENTAL STIMULI

Earthworms exhibit high sensitivity to environmental factors such as moisture, temperature, and light, which significantly influence their behavior and survival. They are highly sensitive to moisture levels and actively seek moist conditions to avoid desiccation, as evidenced by studies on species like *Lumbricus terrestris* (Bouché, 1977). During periods of extreme dryness, earthworms retreat deeper into the soil to escape dehydration, ensuring their survival in challenging conditions (Hendrix & Bohlen, 2002). Temperature also plays a key role in earthworm activity, with optimal temperatures ranging between 10°C and 20°C. Their activity is reduced at both lower and higher temperatures, highlighting their preference for moderate conditions (Edwards, 2004). Furthermore, earthworms are negatively phototropic, meaning they avoid light and remain buried during daylight hours. This behavior serves as a protective mechanism, shielding them from predators such as birds (Denny, 1980).

Together, these sensitivities illustrate how earthworms are finely tuned to their environment, adjusting their behavior to optimize survival.

9.10 REPRODUCTIVE BEHAVIOR

Earthworms exhibit fascinating reproductive behavior, characterized by hermaphroditism, where each individual possesses both male and female reproductive organs. This unique trait allows earthworms to copulate and exchange sperm for fertilization in moist conditions, ensuring successful reproduction (Hartenstein, 1982). After copulation, earthworms produce cocoons, which serve as protective structures where fertilized eggs develop. The number of eggs in each cocoon can vary depending on environmental conditions, such as food availability and soil quality. In some species, greater food resources lead to more eggs being produced per cocoon, demonstrating the influence of environmental factors on reproductive output (Blouin et al., 2013). This reproductive strategy highlights earthworms' adaptability, ensuring their survival and continuity in diverse environmental conditions.

9.11 SOCIAL BEHAVIOR

Earthworms exhibit aggregation behavior, often gathering in areas rich in organic matter or favorable environmental conditions. This aggregation enhances reproductive success by increasing the likelihood of encounters between mates, facilitating higher rates of copulation and fertilization (González et al., 2019). Although earthworms are not typically territorial, they can engage in competition for space and food resources, particularly in environments with high organic content or when populations are dense. In such situations, competition for optimal burrowing sites and nutrient-rich organic matter becomes more intense, as earthworms vie for limited resources (Smith, 2015). This balance of aggregation and competition highlights the complex social dynamics of earthworms, which are crucial for their survival and reproductive success in various ecosystems.

9.12 AVOIDANCE AND DEFENSE BEHAVIOR

Earthworms have evolved several strategies to avoid predation, which is essential for their survival in ecosystems where they face threats from predators such as birds and amphibians. When disturbed, earthworms often retreat deeper into their burrows or burrow more rapidly to escape from predators (Denny, 1980). In addition to behavioral responses, some species of earthworms possess chemical defense mechanisms to deter predators. They secrete substances like slime or noxious chemicals that act as a deterrent, protecting them from potential harm (Gist & Gower, 1987). These strategies, both behavioral and chemical, serve as critical survival tools, enabling earthworms to mitigate the risks posed by natural predators in their environments.

9.13 REACTION TO VIBRATIONS AND SOUNDS

Earthworms are highly sensitive to vibrations in their environment, which often serve as a signal of potential threats from predators or other disturbances. Research by Schmidt and Zangerlé (2001) has shown that earthworms exhibit a clear behavioral response to such vibrations, typically retreating deeper into the soil as a defensive measure. This response helps them avoid predators, such as birds or amphibians, that may detect them through vibrations on the soil surface. By sensing and reacting to these vibrations, earthworms can effectively avoid predation and maintain their safety underground. This ability to detect vibrations reflects the earthworm's acute sensory awareness and its capacity to adapt its behavior to external environmental cues, ensuring its survival in dynamic and often dangerous ecosystems.

9.14 BEHAVIORAL PLASTICITY

The behavioral plasticity of earthworms enables them to adapt to diverse environmental conditions. This adaptability is crucial for their survival, as it allows them to respond to varying factors such as moisture levels and temperature. During periods of drought, when soil moisture is low, earthworms exhibit a behavioral shift by migrating deeper into the soil layers to escape desiccation and avoid dehydration (Edwards & Bohlen, 1996). On the other hand, during flooding events, when the soil becomes waterlogged, earthworms come to the surface to avoid suffocation and drowning. This ability to modify their behavior in response to environmental changes helps ensure their survival and maintain their crucial role in soil health and nutrient cycling. Earthworms' adaptability to both dry and wet conditions underscores their resilience and vital role in ecosystem functioning.

9.15 DISCUSSION

Earthworms exhibit a remarkable range of behaviors that are integral to their ecological functions and survival. Their ability to adapt to varying environmental conditions such as moisture levels, temperature, and soil texture demonstrates their behavioral plasticity and resilience. Through processes like burrowing, feeding on organic matter, and engaging in selective feeding, earthworms contribute significantly to soil aeration, nutrient cycling, and the decomposition of organic materials, enhancing soil fertility. Additionally, their reproductive behaviors, including hermaphroditism and cocoon formation, along with their social dynamics, such as aggregation and competition, further emphasize their critical role in maintaining healthy ecosystems. Earthworms also employ defensive strategies to avoid predators, including retreating into burrows and secreting chemical defenses. Their sensitivity to environmental stimuli, such as vibrations and light, aids in their survival by protecting them from potential threats. Overall, earthworms' behavioral adaptability and ecological contributions underscore their indispensable role in the health and functioning of soil ecosystems, highlighting their importance in both natural and managed environments.

The comparative morphology and anatomy of earthworms illustrate the outstanding integration of external and internal features that enable them to thrive in their ecological niche. Their segmented body and specialized reproductive and digestive systems reflect a finely tuned adaptation to burrowing, nutrient cycling, and reproduction. These structural features not only support the earthworm's survival but also play a vital role in enhancing soil health and fertility, demonstrating the deep connection between the organism's anatomy and its environmental impact.

REFERENCES

Blouin, M., Cordero, N., Ponge, J.-F., & Cluzeau, D. (2013). Influence of environmental factors on earthworm cocoon production. *Soil Biology and Biochemistry*, 63, 97–104.

Bouché, M. B. (1977). Strategies of soil organisms. In *Soil Organisms and Decomposition in the Tropics* (pp. 205–216). Springer.

Darwin, C. (1881). *The Formation of Vegetable Mould through the Action of Worms with Observations on Their Habits*. John Murray.

Denny, M. (1980). Locomotion: The cost of gastropod crawling. *Science*, 208(4447), 1284–1291.

Edwards, C. A. (2004). *Earthworm Ecology: From Darwin to Vermiculture*. CRC Press.

Edwards, C. A., & Bohlen, P. J. (1996). *Biology and Ecology of Earthworms* (3rd ed.). Chapman & Hall.

Gist, E. S., & Gower, J. L. (1987). Chemical defenses in earthworms: Secretion of noxious substances. *Biological Journal of the Linnean Society*, 32(1), 73–81.

González, G., Huang, C. Y., Zheng, X., & Zou, X. (2019). Aggregation behavior of earthworms in response to resource addition in a subtropical earthworm-rich forest. *Pedobiologia*, 72, 1–8. https://doi.org/10.1016/j.pedobi.2018.11.001

Hartenstein, R. (1982). Earthworm mating and cocoon production. *Zoological Journal of the Linnean Society*, 74(2), 273–285.

Hendrix, P. F., & Bohlen, P. J. (2002). Earthworms and soil health. *Nature*, 417, 759–760.

Howmiller, R. P., & Beeton, A. M. (1971). Effects of environmental pollution on earthworm populations. *Environmental Pollution*, 1(2), 113–120.

Julka, J. M. (1988). *Ecology of Earthworms in India*. Indian National Science Academy.

Julka, J. M., & Paliwal, R. (1993). *Ecological Studies on Indian Earthworms: Diversity and Function*. Zoological Survey of India.

Lowe, C. N., & Butt, K. R. (2014). Earthworm responses to soil moisture and temperature. *Soil Biology and Biochemistry*, 76, 137–143.

Parle, J. N. (1962). Burrowing behavior of earthworms and its influence on soil texture. *Journal of Soil Science*, 13(1), 79–92.

Reynolds, J. W., & Wetzel, M. A. (2024). *Earthworm Anatomy and Physiology: A Comparative Approach*. Springer.

Schmidt, O., & Zangerlé, B. (2001). Earthworm behaviour and vibrations: Response to predator disturbances. *Ecology*, 82(4), 1041–1047.

Schulz, H., & Zander, J. (2017). Selective feeding behavior of earthworms: Effects on nutrient cycling. *Biology and Fertility of Soils*, 53(4), 575–582.

Smith, S. (2015). Territoriality and competition among earthworms. *Soil Biology & Biochemistry*, 89, 101–110.

Stephenson, J. (1930). *The Oligochaeta* (978 pp.). Oxford: Clarendon Press.

Zhang, Y., & Xue, S. (2011). Decomposition and nutrient cycling in earthworm-feeding systems. *Soil Biology and Biochemistry*, 43(2), 245–253.

10 Feeding and Digestion of Earthworms

10.1 INTRODUCTION

Earthworms are essential organisms in soil ecosystems, primarily due to their role in breaking down organic matter and recycling nutrients. Their feeding and digestion processes are finely tuned to enhance soil fertility and structure, playing a crucial role in maintaining healthy soils. Earthworms feed on organic material, which they ingest through their mouths and process through a complex digestive system. The digestion of organic matter by earthworms results in nutrient-rich castings, which are beneficial for plant growth and overall soil health (Edwards & Bohlen, 1996). Earthworms are detritivores, feeding on decaying plant material, microorganisms, and other organic matter found in the soil. The digestive process involves both mechanical grinding and biochemical breakdown of food, with the help of enzymes and symbiotic microorganisms within their gut (Suthar, 2009). This process not only helps in breaking down organic material but also contributes to nutrient cycling, making nutrients more available for plants and improving soil structure (Bouché, 1977). The earthworm's role in digestion is vital to sustainable agricultural practices and soil management, as they convert organic waste into valuable nutrients, enhancing the soil's fertility and supporting plant health.

10.2 FEEDING BEHAVIOR OF EARTHWORMS

Earthworms are detritivores, meaning they primarily feed on decomposing organic material such as dead plant matter, decaying animal remains, and other organic detritus. They are opportunistic feeders, capable of consuming a wide variety of organic materials that are available in their environment (Edwards & Bohlen, 1996). Earthworms use their muscular mouthparts, called the prostomium, to grasp organic material and bring it into their mouths. This process is known as "ingestion", and it is often facilitated by mucus produced by specialized cells around the mouth that help bind the organic material (Curry, 2004). Earthworms also feed on microorganisms such as bacteria and fungi, which live on decaying organic matter.

10.3 DIGESTION PROCESS

The digestive process in earthworms is complex and occurs in several stages. Once organic material is ingested, it moves through the earthworm's pharynx into the esophagus and into the crop, a storage area. From there, it travels to the gizzard, where it is mechanically ground up by tiny particles of sand and grit ingested by the earthworm (Suthar, 2009). This grinding process is critical, as earthworms lack teeth and rely on physical abrasion to break down food particles into smaller, digestible

DOI: 10.1201/9781003633600-10

components. The partially digested material then enters the intestine, where enzymes further break down the organic matter into nutrients.

10.4 ABSORPTION AND NUTRIENT CYCLING

In the intestine, digestion continues as enzymes secreted by the earthworm break down organic matter into simpler compounds. Nutrients such as proteins, carbohydrates, and fats are absorbed through the walls of the intestine and into the bloodstream (Bouché, 1977). The undigested matter, primarily composed of minerals and inorganic materials, is excreted as castings, which are rich in nutrients like nitrogen, phosphorus, and potassium. These castings enhance soil fertility, making earthworm activity critical for maintaining healthy, productive soils (Edwards, 2004). Through this process, earthworms contribute to nutrient cycling by converting organic matter into forms that plants can readily absorb.

10.5 THE ROLE OF MICROORGANISMS IN DIGESTION

Earthworms also rely on symbiotic relationships with microorganisms to aid in their digestion. The digestive system of earthworms houses a diverse community of bacteria, fungi, and protozoa, which help decompose organic material. These microorganisms are particularly important for breaking down cellulose and other complex organic compounds that are difficult for earthworms to digest on their own (Barton, 2003). In return, the earthworms provide these microorganisms with a consistent supply of organic material and a stable habitat within their digestive tract.

10.6 DISCUSSION

The feeding and digestion processes of earthworms are vital to soil health and ecosystem functioning. Through their feeding habits, earthworms break down organic matter, recycle nutrients, and contribute to soil aeration and structure. Their digestion process involves both mechanical and biochemical breakdown, aided by the presence of microorganisms in their gut. By processing organic materials into nutrient-rich castings, earthworms enhance soil fertility, making them essential for sustainable agriculture and environmental conservation.

REFERENCES

Barton, D. P. (2003). Earthworm digestion and its ecological significance. *Soil Biology & Biochemistry*, 35(6), 799–809.

Bouché, M. B. (1977). Earthworms: Their ecology and relationships with soils and land use. In *Soil Organisms and Decomposing Organic Matter*. Berlin: Springer.

Curry, J. P. (2004). Feeding and digestion of earthworms: A review of the literature. *Pedobiologia*, 48(1), 43–56.

Edwards, C. A. (2004). *Earthworm Ecology: From Darwin to the Present*. CRC Press.

Edwards, C. A., & Bohlen, P. J. (1996). *Biology and Ecology of Earthworms*. Chapman & Hall.

Suthar, S. (2009). Feeding behavior and digestion in earthworms. *Applied Soil Ecology*, 42(2), 135–141.

11 Studies on Earthworm Gut Microbiome

11.1 INTRODUCTION

Earthworms, grouped under the phylum Annelida, are invertebrates bearing a true coelom (Barnes et al., 2004). By their burrowing activity, they enable mixing of the soil and thus are also termed "farmers' friends". While ingesting soil, organic matter, and litter from soil, and through the metabolism in the earthworm gut, they help in nutrient recycling. Due to their existence in soil, they are frequently exposed to microbial communities. We earlier studied the robust immune system of earthworms in combating pathogens and soil pollutants (Ghosh, 2018; Bilej et al., 2000–2013; Jarosz & Glinski, 1997) that enable their life in soil, and their burrowing activity; ingestion of soil, litter, microbes, and pollutants from soil that undergo various metabolic changes in their gut; and release of casts add to soil fertility and enable maintenance of soil health. Thus, they are also referred to as soil engineers (Phillips et al., 2021).

Because they feed on organic matter, ingesting pollutants and microbes, their gut environment, enzymes and microbiome are of interest. The gut of earthworms has been reported to house diverse microbes, including bacteria, fungi, protozoa, eukaryotes, and ingested microbes that play a role in metabolism, fermentation, and detoxification processes. The gut environment in earthworms has been reported to be anoxic and rich in saccharides that enable anaerobiosis for ingested microbes. In the process of metabolism, they release greenhouse gases (GHG), and recently, only a few reports have highlighted the gut microbiome of earthworms.

In the current chapter, we analyze and review the present state of knowledge regarding the (1) gut microbiome of different species of earthworms under different conditions of soil and feeding, (2) impact of microbiomes under polluted environments, and (3) earthworm gut microbes and generation of GHG.

Different approaches, including amplicon sequencing of bacterial16S rRNA and eukaryotic 18S rRNA from earthworms under *Aporrectodea*, *Allolobophora*, and *Lumbricus* genera, revealed a predominance of bacteria belonging to Acidobacteria, Firmicutes, Actinobacteria, Proteobacteria, and Verrucomicrobia and eukaryotic diversity for the Stramenopiles, Alveolata, and Rhizaria (SAR) supergroup together with the presence of fungi and metazoa. The eukaryotic and prokaryotic communities revealed differences in α-diversity, which was reported to be lowest for *Lumbricus herculeus* (*L. herculeus*), while β-diversity revealed *L. herculeus* housing different microbiomes compared to endogeic earthworms *Aporrectodea caliginosa (A. caliginosa), Allolobophora chlorotica (A. chlorotica)*, and *Aporrectodea tuberculata (A. tuberculata)* and anecic earthworm *Aporrectodea longa (A. longa)*, revealing high interactions between protist and bacterial operational taxonomic

DOI: 10.1201/9781003633600-11

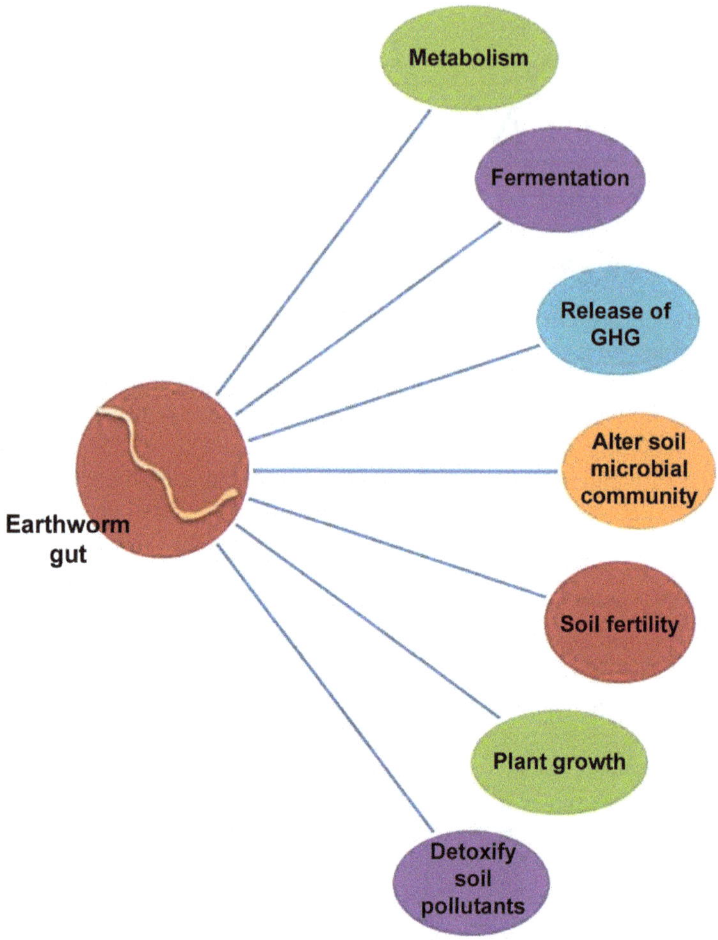

FIGURE 11.1 Earthworm gut microbiome diversity and role.

units (OTUs; Sapkota et al., 2020). Different techniques have revealed the presence of anaerobic ammonium oxidation (anammox) bacteria in the endogeic earthworm *Pheretima guillelmi (P. guillelmi)* gut, of which Brocadia showed the predominant expression (Li et al., 2021a, 2021b; Table 11.1).

Using 16S-rRNA and 16S-rRNA gene sequences, studies on the anecic, deep-burrowing *Lumbricus terrestris* (*L. terrestris)* gut revealed the exclusive occurrence and activity of dominant forms of Proteobacteria and Firmicutes phylotypes not reported from soil (Zeibich et al., 2019a, 2019b, 2019c).

In the process of feeding and releasing casts, endogeic and anecic earthworms have been reported to add fertility to the soil, alter the microbial communities in soil, and cause nitrogen recycling, although the exact processes involved are far from being completely understood (Medina-Sauza et al., 2019).

11.2 EARTHWORM MICROBIOTA AND METABOLISM

In earthworms, metabolism of organic matter occurs in the foregut and midgut, and it is excreted by the hindgut. Earthworm gut microbiome contents have been studied by analysisng the bacterial 16S rRNA expression and analysed by the FAPRATOX database that takes into account the genera and taxa of microbes involved in functions including nitrification, denitrification, fermentation, revealed reduced dominant phyla Bacteroidetes and Proteobacteria but increase in Actinobacteria, Acidobacteria and Chloroflexi expression during transit through gut. With the lowest bacterial alpha diversity of the midgut, differences in bacterial communities were reported from the foregut with that of the mid- and hindgut. Microbes metabolizing nitrogen (N) and carbon (C) revealed predominant expression in the gut, and those involved in methanogenesis and fermentation were predominant in the hindgut, with microbes involved in the denitrification process predominant in the foregut. (Medina-Sauza et al., 2019; Wang et al., 2021). It was also observed that while microbes involved in N metabolism in vermicompost increased during gut transit, the pathogenic microbes revealed inhibition in growth (Medina-Sauza et al., 2019; Wang et al., 2021).

While ingesting soil organic matter and with its passage through the gizzard and gut, the contents are disrupted, and RNA and protein content is released from disrupted microbial cells, facilitating fermentations reaction by ingested bacteria. Amino acids, including aspartate, glutamate, and threonine, have been reported to stimulate fermentation in the *L. terrestris* gut by different groups of bacteria analyzed by 16S rRNA studies. Glutamate has been reported to stimulate *Fusobacteriaceae*, leading to the formation of H_2 and formate. Aspartate has been reported to stimulate *Aeromonadaceae*, leading to the formation of succinate and propionate.

L. terrestris gut microbiota has been reported to ferment starch, glycogen stimulating production of H_2, ethanol, fatty acid formation and revealed presence of facultative fermentative microbial families including Aeromonadaceae and Enterobacteriaceae by 16S rRNA based sequencing approaches.

The *L. terrestris* gut revealed activity of different obligate anaerobes and facultative aerobes involved in fermentation reactions. While protein stimulated activity of *Clostridiaceae,* the *Fusobacteriaceae* and *Peptostreptococcaceae* group of bacteria RNA stimulated activity of *Aeromonadaceae*, leading to increased organic carbon (C) in the earthworm gut. Earthworms ingest microbes together with contaminants like biopolymers from soil that undergo fermentation in anoxic environments in the gut and generate fatty acids (Zeibich et al., 2018).

Saccharides, including arabinose, galactose, glucose, fucose, maltose, mannose, rhamnose, and xylose, have been reported to undergo fermentation by *L. terrestris* gut microbes, producing different organic acids, acetate, CO_2, ethanol, formate H_2, lactate, propionate, and succinate, wherein the dominant facultative aerobes include *Aeromonadaceae* and *Enterobacteriaceae* (Meier et al., 2018).

Vermicomposting by earthworms causes a reduction in pathogens, reducing harmful effects of waste, and the process is selective based on the Gram-positive or Gram-negative nature, due to differences in their cell wall composition (Swati &

Hait, 2018). Vermicompost produced from organic wastes revealed increased humic acid, which was revealed to play a role in plant growth (Bhat et al., 2018).

The fermentation process in the epigeic earthworm *Eudrilus eugeniae (E. eugeniae)* gut has been reported to emit methane (CH_4) by the activity of ingested methanogens. Glucose has been reported to undergo fermentation to produce acetate, carbon dioxide (CO_2), H_2, ethanol, formate, lactate, propionate, and succinate, mediated by the activity of facultative aerobes; Aeromonadaceae; and obligate anaerobes, including Lachnospiraceae, Ruminococcaceae, and Veillonellaceae. Hydrogenotrophic methanogens including Methanobacteriaceae and Methanoregulaceae and acetogens including Peptostreptoccocaceae have been reported to contribute to methanogenesis and to the methanogenic food web (Schulz et al., 2015). *E. eugienea*, when fed with a rich organic diet, showed increased biomass with high bacterial count and the presence of Bacillus, Lactobacillus, and Flavobacterium and release of enzymes in the earthworm gut, including amylase, cellulase, endoglucanase, protease, and sucrase (Sumathi & Thaddeus, 2013).

The earthworm species *Eisenia fetida (E. fetida), Eisenia andrei (E. andrei), Eisenia eugeniae (E. eugeniae*) and anecic *Octodrilus complanatus (O. complanatus)* were fed pig slurry to study its passage through gut. It revealed differences in bacterial counts that differed between the species. Although all species could degrade pig slurry and no differences were reported in their gut dehydrogenase enzyme activity, the bacterial count of *E. fetida* and *E. eugeniae* gut content remained unaltered, while it reduced for *O. complanatus* and *E. andrei* (Aira et al., 2009).

Diverse forms of actinomycetes, bacteria, fungi, protozoa, and yeast have been reported from casts and guts of epigeic species *E. eugeniae*, *E. fetida*, and *Perionyx excavatus (P. excavatus)* and the anecic earthworm *Lampito mauritii (L. mauritii)*, influenced by feed. While Actinomycetes including *Nocardia asteroides (N. asteroides), N. caviae, Streptomyces albus (S. albus), Streptomyces somaliensis (S. somaliensis)*, and species under genera *Saccharomonosporia* remained undigested, protozoa including *Amoeba proteus (A. proteus), Apospathidium terricola (A. terricola), Euglena viridis (E. viridis), E. orientalis, Paramecium trichium (P. trichium), Trichomonas hominis (T. hominis), Vorticella picta (V. picta)* and yeast including *Candida albicans (C. albicans), C. krusei, C. tropicalis*, and *Cryptococcus neoformans (C. neoformans)* were reported to undergo digestion.

While Fungi including *Alternaria sp., Cladosporium carrionii (C. carrionii), mold, Curvalaria lunata (C. lunata), Curvularia geniculata (C. geniculata), Cunninghamella echinulata (C. echinulata), Geotrichum candidum (G. candidum), Micrococci, Mucor plumbeus (M. plumbeus), Mycetia sterile (M. sterile), Saksenae vasiformis (S. vasiformis), Syncephalostrum racemosum (S. racemosum)* and bacteria including Gram positive, *Bacillus subtilis (Bacillus subtilis), Bacillus cereus (B. cereus), Bacterium antitratum (B. antitratum), Enterococci species, Enterobacter aerogenes (E. aerogenes), gram negative Escherichia coli (E. coli), Enterobacter cloacae (E. cloacae), Mima polymorpha (M. polymorpha), Pseudomonas aeruginosa (P. aeruginosa), Proteus mirabilis (P. mirabilis), P. rettgeri, P. vulgaris, Staphylococus citreus (S. citreus)* revealed complete digestion, Fungi including *Aspergillus flavus (A. flavus), A fumigatus, A. ochraceous and Rhizopus sp. Klebsiella pneumoniae (K. pneumoniae) and Morganella morganii (M. morganii)* bacteria remained undigested during passage through the gut. *Trichoderma koningii* (*T. koningii*) and *Fusarium moniliforme (F. moniliforme)* were not digested by any species other than *E. eugeniae.* The earthworms reported increased microbial proliferation in the casts and caused increased microbial activity, with increased humic acid contents in the casts (Parthasarathi et al., 2007).

The microbes in the gut wall of *A. caliginosa* have been reported to produce nitrite and nitrous oxide (N_2O) under anoxic conditions studied by 16S rRNA–based studies (Ihssen et al., 2003). Aerobic and anaerobic gut microbes of epi-endogeic species *Lumbricus rubellus (L. rubellus)* and endogeic *Octolasium lacteum (O. lacteum)* and gut homogenates have been reported to metabolize glucose, cellobiose, or ferulate under anaerobic conditions and consume cellobiose or oxygen under aerobic conditions (Karsten & Drake, 1995).

11.3 EARTHWORM GUT MICROBIOMES AND POLLUTANTS

Polymyxin B (PMB) combined with arsenic (As) as pollutants to which earthworms were exposed affected earthworm gut microbial composition, causing upregulation of antibiotic resistance genes (ARGs) and mobile genetic elements (MGEs). PMB treatment increased Actinobacteria expression, while As stress increased expression of Proteobacteria in the gut (Li et al., 2021a, 2021b). Environmental pollutants like biocidal nanoparticles (NPs), including Ag and CuO, affect the gut microbiome of the *E. fetida* (Swart et al., 2020) earthworm. The organophosphorus pesticide monocrotophos could reduce proliferation and degradation of *L. mauritii* gut microbiota together with altered histopathology, degenerated nuclei, damaged villi, pathology of vacuolization, congestion in intestinal blood sinuses, degradation of fungi, and reduced regenerative potential of intestinal chloragogen cells (Kavitha et al., 2020). Bioaccumulated perfluorooctane sulfonamide (FOSA) as a soil pollutant could be biotransformed to perfluorooctane sulfonate (PFOS) by enzymes and gut microbes in earthworms (Zhao et al., 2020). The soil contaminant arsenic, as arsenite (As(III)), also found to be predominant in the earthworm gut, has been reported to undergo biotransformation by arsenic biotransformation genes (ABGs) by gut microbiota in *Metaphire sieboldi (M. sieboldi,* Wang et al., 2019). Microplastic pollutants fed to *Enchytraeus crypticus (E. crypticus)* have been reported to affect its growth and gut microbiota. 16S rRNA amplification revealed a reduced relative abundance of microbes of Isosphaeraceae, Rhizobiaceae, and Xanthobacteraceae families, playing a role in the degradation of organic matter and N cycling (Zhu et al., 2018). Earthworm chloragocyte cells and gut microbes, together with the metabolic processes in their gut, have been reported to detoxify heavy metal pollutants from industrial sources (Bhat et al., 2018).

The soil and groundwater contaminant 2,4-dichlorophenol (2,4-DCP) can be degraded by soil microbes, which were also found in the gut of the endogeic earthworm *A. caliginosa*, as revealed by comparative amplicon 16S rRNA–based pyrosequencing studies. Comamonas, Desulfitobacterium, and Novosphingobium have been reported to assimilate radioactive 2,4-DCP-[(13)C] even without the presence of earthworms. Clostridium, Flavobacterium, and Pseudomonas have been reported to assimilate 2,4-DCP-[(13)C] in the drilosphere, while Novosphingobium and Variovorax were reported to be predominant in soil slurries in the drilosphere (Dallinger & Horn, 2014, Table 11.1).

P. carnosa earthworm gut bacteria has been reported to cause weathering of silicate minerals. Amplified ribosomal DNA restriction analysis (ARDRA) revealed increased bacterial diversity in the guts of earthworms that were fed with potassium-bearing rock powder (PBRP; Liu et al., 2011). Ergovaline (EV) is an alkaloid that can cause disease in livestock. Thus, microbes that can detoxify EV are important. EV-degrading microbes have been reported from the anaerobic gut of *E. fetida* by 16S rRNA–based studies (Rattray et al., 2010; Table 11.1).

11.4 EARTHWORM GUT MICROBES AND GHG

The anoxic gut environment of the earthworm gut has been revealed to enable housing and survival of denitrifying and fermentative bacteria ingested as soil microbes, which have been reported to release dinitrogen (N_2) and N_2O, which is a greenhouse gas affecting the culturability, fitness, and diversity of diverse soil microbial biomes (Drake & Horn, 2007). The geophagous endogeic earthworm *P. guillelmi* has been reported to emit N_2 and N_2O with the action of the anammox bacterial community in its gut (Li et al., 2021a, 2021b).

11.5 WILDLIFE AND THE SCIENCE OF CONSERVATION AND EARTHWORM RESEARCH

Despite the known importance of microbial symbionts affecting hosts in many ways, modern day research is focused on conservation biology on host–microbiome associations for endangered species primarily due to anthropogenic causes. Host–microbiomes studies have highlighted the role of the microbiome's importance from the viewpoint of conservation, but such studies are mostly based on vertebrates, like the alpine musk deer (Sun et al., 2020) and Amur tiger (Ning et al., 2020). Environmental conditions find importance in shaping host–microbiome associations and have been reported from the microbiome of the Chinese giant salamander (Feng et al., 2022). Co-evolution of marsupials and their gut microbiomes with predominance of, Bacteroidetes, Firmicutes, and Proteobacteria has been studied (Zhu et al., 2021).

Such studies have not yet been reported from the earthworm and largely need to be carried out to understand the role of the host–microbiome relationship in the conservation of rare or endangered species or reintroduction of a species in hitherto-unreported zones. Large-scale use of metagenomic studies with bioinformatic tool-based analysis (Dong et al., 2020) is needed.

TABLE 11.1
Earthworm Gut Microbial Diversity and Roles

Gut content					
Species of Earthworm		**Microbes**	**Method of Detection**	**Role**	**Ref**
A. caliginosa, A. longa, A. tuberculata, A. chlorotica, and *L. herculeus*		• Acidobacteria • Actinobacteria • Firmicutes • Proteobacteria • Verrucomicrobia • SAR supergroup • Fungi • Metazoa	• Amplicon sequencing • 16S rRNA for bacteria • 18S rRNA for eukaryotes	Gut metabolism	Sapkota et al., 2020
P. guillelmi		Anammox bacteria	• 15N-tracing • Quantitative PCR • Bacterial 16S rRNA amplicon sequencing	• Gut metabolism • Release of N_2, N_2O	Li et al., 2021a, 2021b
L. terrestris		• Proteobacteria • Firmicutes	16S rRNA	Fermentation	Zeibich et al., 2019a
Feed	**Component/substrates**				
E. fetida	Feed with composted cow manure	• Reduced Proteobacteria and Bacteroidetes • Increased Actinobacteria, Acidobacteria, and Chloroflexi • Predominant C- and N-metabolizing microbes	• Bacterial 16S rRNA • Biochemical analyses • Analyses based on FAPROTAX databases	• Hindgut: fermentation and methanogenesis • Foregut: denitrification, mineralization of organic matter and phosphorus and potassium enrichment	Wang et al., 2021

(Continued)

TABLE 11.1 (*Continued*)
Earthworm Gut Microbial Diversity and Roles

Species of Earthworm	Gut content	Microbes	Method of Detection	Role	Ref
L. terrestris	• Aspartate • Glutamate • Threonine	• Glutamate stimulated *Fusobacteriaceae*, yielding H_2 and formate • Aspartate stimulated *Aeromonadaceae* and yielded succinate and propionate • *Peptostreptococcaceae* and *Lachnospiraceae* caused Stickland fermentation • Ribose fermentation facilitated by *Aeromonadaceae*	16S rRNA-based methods	Fermentation, Amino acids and ribose enable fermentation process	Zeibich et al., 2019c
L. terrestris	• Cellulose, chitin undergo fermentation • Starch, glycogen stimulate H_2, ethanol, fatty acid formation • Glucose, GlcNAc	• Facultative fermentative families Aeromonadaceae, Enterobacteriaceae, and Fusobacteriaceae	16S rRNA based methods	Obligate anaerobes and facultative aerobes involved in fermentation	Zeibich et al., 2019b
L. terrestris	• Protein stimulated *Peptostreptococcaceae*, *Clostridiaceae*, and *Fusobacteriaceae* • RNA stimulated *Aeromonadaceae*	Fermentation in anoxic environment increases C content in gut and generates fatty acids	16S rRNA and 16S rRNA gene analyses	Obligate anaerobes and facultative aerobes involved in fermentation	Zeibich et al., 2018
L. terrestris	Arabinose, fucose galactose, glucose, maltose, mannose, rhamnose, and xylose stimulate acetate, CO_2, ethanol, formate, H_2, lactate, propionate, succinate production	Fermentation by dominant facultative aerobes *Aeromonadaceae* and *Enterobacteriaceae*	16S rRNA–based methods	Fermentation	Meier et al., 2018

TABLE 11.1 *(Continued)*
Earthworm Gut Microbial Diversity and Roles

Species of Earthworm	Gut content	Microbes	Method of Detection	Role	Ref
E. eugeniae	Saccharide-rich gut	• Glucose is fermented by aeromonadaceae facultative aerobes and obligate anaerobes Lachnospiraceae, Ruminococcaceae, Veillonellaceae. • Methanogens contribute to methanogenesis Acetogens contribute to methanogenic food web	• Bacterial 16S rRNA • alpha subunit of methyl-CoM reductase activity • Methanogen isoenzymes	• Methanogenesis • CH_4 emitted by methanogens	Schulz et al., 2015
E. eugienea	Organic rich diet	Caused high bacterial count and presence of *Bacillus* sp., *Lactobacillus* sp., and *Flavobacterium* sp.	• 16S rRNA–based methods	• Metabolism • Release of amylase, endoglucanase, cellulase, sucrase, and protease in gut • Increased biomass	Sumathi & Thaddeus, 2013
E. fetida, E andrei, E. eugeniae, O. complanatus	Fed with pig slurry	Reduced bacterial count in *O. complanatus* and *E. andrei* while unchanged in *E. fetida* and *E. eugeniae*	Analytical methods	Degraded pig slurry	Aira et al., 2009

(Continued)

TABLE 11.1 *(Continued)*
Earthworm Gut Microbial Diversity and Roles

Species of Earthworm	Gut content	Microbes	Method of Detection	Role	Ref
E. eugeniae, L. mauritii, E. fetida, and P. excavatus	Cow dung, clay loam soil, and mud	• Actinomycetes including *N. asteroides, N. caviae, Streptomyces albus, S. somaliensis,* and *Saccharomonospora* not digested • *A. proteus, A. terricola, E. viridis, E. orientalis, P. trichium, T. hominis,* and *V. picta* and yeast including *C tropicalis, C. krusei, C. albicans,* and *C. neoformans* underwent digestion • Fungi including *Alternaria, C. carrionii, C. herbacium, C. lunata, C. geniculata, C. echinulata, G. candidum Micrococci, Mucor plumbeus, Mycetia sterile, Saksenae vasiformis,* and *S. racemosum* and bacteria including *B subtilis, B. cereus, Bantitratum, Enterococci, E. coli, E. aerogenes, E. cloacae, M. polymorpha, P. aeruginosa, P. vulgaris, P. mirabilis, P. rettgeri,* and *S. citreus* revealed complete digestion • Fungi: *A fumigatus, A. flavus, A. ochraceous*, and Rhizopus and bacteria: *K. pneumoniae* and *M morganii* were not digested but grew in numbers during passage through the gut • *T. koningii* and *F. moniliforme* were not digested by any species other than *E. eugeniae*	• Sequencing-based methods	Metabolism, growth	Parthasarathi et al., 2007
A. caliginosa	Soil	Microbes in gut	• 16S rRNA gene sequence–based identification • Detection for nitrate- or nitrite-reducing bacteria isolated from the gut	Release of nitrite and N_2O under anoxic conditions	Ihssen et al., 2003

(Continued)

TABLE 11.1 *(Continued)*
Earthworm Gut Microbial Diversity and Roles

Species of Earthworm		Microbes	Method of Detection	Role	Ref
Gut content					
L. rubellus Hoffmeister and O. lacteum	• Chromic luvisol soil from limestone as parent material • Silty loam soil	Microbes in gut	Analytical methods to detect the metabolites released	• Metabolism of glucose, cellobiose or ferulate under anaerobic conditions • Consume cellobiose or oxygen under aerobic conditions	Karsten & Drake, 1995
Earthworm	Ingested soil	Denitrifying and fermentative bacteria	Microbial staining methods	• Denitrification and fermentation • Release of GHG: N_2 and N_2O	Drake & Horn, 2007
Pollutants					
M. guillelmi	Pollutant polymyxin B (PMB) with arsenic (As)	• Affected microbial population • Upregulated MGE, ARG • PMB increased Actinobacteria As increased Proteobacteria	Biochemical, microbial studies	Resilient to pollutants	Li et al., 2021a, 2021b
E. fetida	Ag and CuO biocidal NPs	• Affected gut microbial population • Reduced *Candidatus Lumbricincola*	• Statistical analysis: multivariate, differential abundance, species sensitivity distribution analysis	Gut microbiome resilient to biocidal NPs	Swart et al., 2020

(Continued)

TABLE 11.1 *(Continued)*
Earthworm Gut Microbial Diversity and Roles

Gut content

Species of Earthworm		Microbes	Method of Detection	Role	Ref
L. mauritii	Organophosphorus pesticide monocrotophos	• Reduced proliferation of gut microbiota • Altered histopathology, vacuolization • Degenerated nuclei • Damaged villi • Congested intestinal blood sinuses	Microbial methods and analytical methods	Degrade bacteria and fungi and reduce the regenerative potential of intestinal chloragogen cells	Kavitha et al., 2020
E. fetida	FOSA metabolized to PFOS	Enzymes and gut microbes	Microbial and analytical methods	Metabolism	Zhao et al., 2020
M. sieboldi	Arsenic as arsenite contaminant	Biotransformation of arsenic as arsenite by arsenic biotransformation genes by gut microbiota	Microbial and analytical methods	Biotransformation	Wang et al., 2019
E. crypticus	Microplastic contamination affected gut microbiota	Rhizobiaceae, Xanthobacteraceae, and Isosphaeraceae	16S rRNA amplification	Nitrogen cycling and organic matter decomposition	Zhu et al., 2018
A. caliginosa	2,4-DCP–contaminated soil	Comamonas, Desulfitobacterium, and Novosphingobium	Amplicon 16S rRNA-based pyrosequencing	Assimilate 2,4-dichlorophenol (2,4-DCP)	Dallinger & Horn, 2014
P. carnosa	Earthworm fed with PBRP, silicate mineral	*Gammaproteobacteria*-dominant organism	• Transmission electron microscope (TEM) • Amplified ribosomal DNA restriction analysis (ARDRA) • 16S rRNA–based studies for bacteria	• Weathering of silicate mineral • Release of HNO_3, Al, Fe, and Ca	Liu et al., 2011
E. fetida	Treated with EV	Gut microbes	16S-rRNA–based studies	Degrade EV	Rattray et al., 2010

11.6 DISCUSSION

The gut microbiota of invertebrates inhabiting the soil has been reported to harbor microbes, including Bacteria, Archaea, and Eukarya. The intestines have been reported to house diverse eukaryotes, including protozoa, yeasts and fungi, proteobacteria including Gram-positive bacteria, Actinobacteria, Bacteroides, Firmicutes, Flavobacterium, spirochetes, and methanogens under Archaea. Intestinal bacteria have been reported to degrade aromatic compounds, cellulose, and hemicelluloses, and Bacilli has been found to be predominantly involved in cellulose degradation. Nitrogen fixation by microbes has been reported to contribute to the redox environment of the gut, thereby contributing to the generation of GHGs (König, 2006; Table 11.1).

Recent studies have highlighted that the release of GHGs like N compounds and methane (CH_4) by earthworms may pose an environmental hazard, as it is hypothesized to increase the GHG content in the environment. Thus the role of earthworms as friend or foe remains questionable (Schulz et al., 2015; Lubbers et al., 2013). Contrasting reports also exist on earthworms' role in increasing soil fertility, altering the soil microbiota, and detoxification of pollutants (Figure 11.1). However, the current state of the art reveals knowledge gaps as to the assessment of the exact role of earthworms in contributing to the advantage or disadvantage of the environment, highlighting the importance of more research in this domain of life sciences.

Not many studies have been conducted along these lines on earthworm species that are reported from India. It is therefore our next endeavor in understanding the earthworm gut microbiota in the context of soil in India and their relation to agriculture.

REFERENCES

Aira, M., Monroy, F., & Domínguez, J. (2009). Changes in bacterial numbers and microbial activity of pig slurry during gut transit of epigeic and anecic earthworms. *Journal of Hazardous Materials*, 162(2–3), 1404–1407. https://doi.org/10.1016/j.jhazmat.2008.06.042

Barnes, R. D., Edward, E., Fox, R. S., & Barnes, R. D. (2004). *Invertebrate Zoology: A Functional Evolutionary Approach* (7th ed.). Belmont, CA: Thomson-Brooks/Cole.

Bhat, S. A., Singh, S., Singh, J., Kumar, S., Bhawana, & Vig, A. P. (2018). Bioremediation and detoxification of industrial wastes by earthworms: Vermicompost as powerful crop nutrient in sustainable agriculture. *Bioresource Technology*, 252, 172–179. https://doi.org/10.1016/j.biortech.2017.12.013

Bilej, M., Procházková, P., Šilerová, M., & Josková, R. (2011). Earthworm immunity. In *Landes Bioscience, Madame Curie Bioscience Database*. Landes Bioscience. Available from https://www.ncbi.nlm.nih.gov/books/NBK6274/ (Accessed 26 October 2023).

Dallinger, A., & Horn, M. A. (2014). Agricultural soil and drilosphere as reservoirs of new and unusual assimilators of 2,4-dichlorophenol carbon. *Environmental Microbiology*, 16(1), 84–100. https://doi.org/10.1111/1462-2920.12134

Dong, J., Liu, S., Zhang, Y., Dai, Y., & Wu, Q. (2020). A new alignment-free whole metagenome comparison tool and its application on gut microbiomes of wild giant pandas. *Frontiers in Microbiology*, 11, 1061. https://doi.org/10.3389/fmicb.2020.01061

Drake, H. L., & Horn, M. A. (2007). As the worm turns: The earthworm gut as a transient habitat for soil microbial biomes. *Annual Review of Microbiology*, 61, 169–189. https://doi.org/10.1146/annurev.micro.61.080706.093139

Feng, J., Zhu, W., Jiang, J., Zhao, C., Sun, Z., Jiang, W., Luo, Q., & Zhao, T. (2022). Reintroduction modifies the intraspecific variations of symbiotic microbes in captive bred Chinese giant salamander. *Frontiers in Microbiology*, 13, 1062604. https://doi.org/10.3389/fmicb.2022.1062604

Ghosh, S. (2018). Environmental pollutants, pathogens and immune system in earthworms. *Environmental Science and Pollution Research*, 25(7), 6196–6208. https://doi.org/10.1007/s11356-017-0825-2

Ihssen, J., Horn, M. A., Matthies, C., Gössner, A., Schramm, A., & Drake, H. L. (2003). N_2O-producing microorganisms in the gut of the earthworm *Aporrectodea caliginosa* are indicative of ingested soil bacteria. *Applied and Environmental Microbiology*, 69(3), 1655–1661. https://doi.org/10.1128/AEM.69.3.1655-1661.2003

Jarosz, J., & Gliński, Z. (1997). Earthworm immunity: The basic mechanisms. *Folia Biologica (Kraków)*, 45(3–4), 143–154.

Karsten, G. R., & Drake, H. L. (1995). Comparative assessment of the aerobic and anaerobic microfloras of earthworm guts and forest soils. *Applied and Environmental Microbiology*, 61(3), 1039–1044.

Kavitha, V., Anandhan, R., Alharbi, N. S., Kadaikunnan, S., Khaled, J. M., Almanaa, T. N., & Govindarajan, M. (2020). Impact of pesticide monocrotophos on microbial populations and histology of intestine in the Indian earthworm *Lampito mauritii* (Kinberg). *Microbial Pathogenesis*, 139, 103893. https://doi.org/10.1016/j.micpath.2019.103893

König, H. (2006). *Bacillus* species in the intestine of termites and other soil invertebrates. *Journal of Applied Microbiology*, 101(3), 620–627. https://doi.org/10.1111/j.1365-2672.2006.02939.x

Li, H., Yang, X. R., Wang, J., Zhou, G. W., Zhang, Y. S., Lassen, S. B., Zhu, Y. G., & Su, J. Q. (2021a). Earthworm gut: An overlooked niche for anaerobic ammonium oxidation in agricultural soil. *Science of the Total Environment*, 752, 141874. https://doi.org/10.1016/j.scitotenv.2020.141874

Li, L., Zhu, D., Yi, X., Su, J., Duan, G., Tang, X., & Zhu, Y. (2021b). Combined pollution of arsenic and Polymyxin B enhanced arsenic toxicity and enriched ARG abundance in soil and earthworm gut microbiotas. *Journal of Environmental Sciences (China)*, 109, 171–180. https://doi.org/10.1016/j.jes.2021.01.019

Liu, D., Lian, B., Wang, B., & Jiang, G. (2011). Degradation of potassium rock by earthworms and responses of bacterial communities in its gut and surrounding substrates after being fed with mineral. *PLoS ONE*, 6(12), e28803. https://doi.org/10.1371/journal.pone.0028803

Lubbers, I. M., van Groenigen, K. J., Fonte, S. J., Six, J., Brussaard, L., & van Groenigen, J. W. (2013). Greenhouse-gas emissions from soils increased by earthworms. *Nature Climate Change*, 3(2), 187–194. https://doi.org/10.1038/nclimate1692

Medina-Sauza, R. M., Álvarez-Jiménez, M., Delhal, A., Reverchon, F., Blouin, M., Guerrero-Analco, J. A., Barois, I., & Sánchez-Rodríguez, L. M. (2019). Earthworms building up soil microbiota: A review. *Frontiers in Environmental Science*, 7, 81. https://doi.org/10.3389/fenvs.2019.00081

Meier, A. B., Hunger, S., & Drake, H. L. (2018). Differential engagement of fermentative taxa in gut contents of the earthworm *Lumbricus terrestris*. *Applied and Environmental Microbiology*, 84(5), e01851-17. https://doi.org/10.1128/AEM.01851-17

Ning, Y., Qi, J., Dobbins, M. T., Liang, L., Wang, C., Li, S., & Wei, J. (2020). Comparative analysis of microbial community structure and function in the gut of wild and captive Amur tiger. *Frontiers in Microbiology*, 11, 1665. https://doi.org/10.3389/fmicb.2020.01665

Parthasarathi, K., Ranganathan, L. S., Anandi, V., & Zeyer, J. (2007). Diversity of microflora in the gut and casts of tropical composting earthworms reared on different substrates. *Journal of Environmental Biology*, 28(1), 87–97.

Phillips, H. R. P., Bach, E. M., Bartz, M. L. C., Bennett, J. M., Beugnon, R., Briones, M. J. I., Brown, G. G., Ferlian, O., Gongalsky, K. B., Guerra, C. A., König-Ries, B., Krebs, J. J., Orgiazzi, A., Ramirez, K. S., Russell, D. J., Schwarz, B., Wall, D. H., & Eisenhauer, N. (2021). Global data on earthworm abundance, biomass, diversity and corresponding environmental properties. *Scientific Data*, 8(1), 136. https://doi.org/10.1038/s41597-021-00907-y

Rattray, R. M., Perumbakkam, S., Smith, F., & Craig, A. M. (2010). Microbiomic comparison of the intestine of the earthworm *Eisenia fetida* fed ergovaline. *Current Microbiology*, 60(3), 229–235. https://doi.org/10.1007/s00284-009-9538-z

Sapkota, R., Santos, S., Farias, P., Krogh, P. H., & Winding, A. (2020). Insights into the earthworm gut multi-kingdom microbial communities. *Science of the Total Environment*, 727, 138301. https://doi.org/10.1016/j.scitotenv.2020.138301

Schulz, K., Hunger, S., Brown, G. G., Tsai, S. M., Cerri, C. C., Conrad, R., & Drake, H. L. (2015). Methanogenic food web in the gut contents of methane-emitting earthworm *Eudrilus eugeniae* from Brazil. *ISME Journal*, 9(8), 1778–1792. https://doi.org/10.1038/ismej.2014.266

Sumathi, G., & Thaddeus, A. (2013). Impact of organic rich diet on gut enzymes, microbes and biomass of earthworm, *Eudrilus eugienea*. *Journal of Environmental Biology*, 34(3), 515–520.

Sun, Y., Shi, Z., Liu, Z., Zhao, C., Huang, X., Li, Q., & Huang, J. (2020). Gut microbiota of wild and captive Alpine musk deer (Moschus chrysogaster). *Frontiers in Microbiology*, 10, 3156.

Swart, E., Goodall, T., Kille, P., Spurgeon, D. J., & Svendsen, C. (2020). The earthworm microbiome is resilient to exposure to biocidal metal nanoparticles. *Environmental Pollution*, 267, 115633. https://doi.org/10.1016/j.envpol.2020.115633

Swati, A., & Hait, S. (2018). A comprehensive review of the fate of pathogens during vermicomposting of organic wastes. *Journal of Environmental Quality*, 47(1), 16–29. https://doi.org/10.2134/jeq2017.03.0127

Wang, H. T., Zhu, D., Li, G., Zheng, F., Ding, J., O'Connor, P. J., Zhu, Y. G., & Xue, X. M. (2019). Effects of arsenic on gut microbiota and its biotransformation genes in earthworm *Metaphire sieboldi*. *Environmental Science & Technology*, 53(7), 3841–3849. https://doi.org/10.1021/acs.est.8b05952

Wang, N., Wang, W., Jiang, Y., Dai, W., Li, P., Yao, D., & Jiang, S. (2021). Variations in bacterial taxonomic profiles and potential functions in response to the gut transit of earthworms (Eisenia fetida) feeding on cow manure. *Science of the Total Environment*, 787, 147392. https://doi.org/10.1016/j.scitotenv.2021.147392

Zeibich, L., Schmidt, O., & Drake, H. L. (2018). Protein- and RNA-enhanced fermentation by gut microbiota of the earthworm *Lumbricus terrestris*. *Applied and Environmental Microbiology*, 84(11), e00657-18. https://doi.org/10.1128/AEM.00657-18

Zeibich, L., Schmidt, O., & Drake, H. L. (2019a). Fermenters in the earthworm gut: Do transients matter? *FEMS Microbiology Ecology*, 95(2), fiz001. https://doi.org/10.1093/femsec/fiz001

Zeibich, L., Schmidt, O., & Drake, H. L. (2019b). Dietary polysaccharides: Fermentation potentials of a primitive gut ecosystem. *Environmental Microbiology*, 21(4), 1436–1451. https://doi.org/10.1111/1462-2920.14552

Zeibich, L., Staege, M., Schmidt, O., & Drake, H. L. (2019c). Amino acids and ribose: Drivers of protein and RNA fermentation by ingested bacteria of a primitive gut ecosystem. *Applied and Environmental Microbiology*, 85(19), e01297-19. https://doi.org/10.1128/AEM.01297-19

Zhao, S., Wang, B., Zhong, Z., Liu, T., Liang, T., & Zhan, J. (2020). Contributions of enzymes and gut microbes to biotransformation of perfluorooctane sulfonamide in earthworms (*Eisenia fetida*). *Chemosphere*, 238, 124619. https://doi.org/10.1016/j.chemosphere.2019.124619

Zhu, B. K., Fang, Y. M., Zhu, D., Christie, P., Ke, X., & Zhu, Y. G. (2018). Exposure to nanoplastics disturbs the gut microbiome in the soil oligochaete *Enchytraeus crypticus*. *Environmental Pollution*, 239, 408–415. https://doi.org/10.1016/j.envpol.2018.04.088

Zhu, L., Wang, J., & Bahrndorff, S. (2021). Editorial: The wildlife gut microbiome and its implication for conservation biology. *Frontiers in Microbiology*, 12, 697499. https://doi.org/10.3389/fmicb.2021.697499

12 Economic Importance of Earthworms

12.1 INTRODUCTION

Earthworms, commonly known as "nature's plow", are vital for maintaining soil fertility and enhancing agricultural productivity. Their biological activities, such as burrowing and feeding on organic matter, improve soil structure, enhance nutrient cycling, and contribute to the health of ecosystems. These humble invertebrates, through their natural processes, have become a cornerstone of sustainable farming, environmental conservation, and waste management. Their ability to recycle organic waste into valuable products like vermicompost and their role in promoting soil aeration make them invaluable to modern agricultural practices (Edwards & Bohlen, 1996). In addition to their ecological importance, earthworms have significant economic value. They are central to the burgeoning industry of vermiculture, which involves the cultivation of earthworms for the production of high-quality compost, animal feed, and even as bait for fishing (Binet et al., 1998). Vermiculture offers a sustainable method of organic waste disposal and enhances soil quality, reducing dependency on chemical fertilizers and pesticides (Suthar, 2009). The global vermiculture industry has grown in recent decades, with earthworm-based products increasingly used in organic farming, horticulture, and waste management (Vargas et al., 2019). The economic impact of earthworms extends beyond agriculture into various other sectors. They contribute to environmental sustainability, particularly in bioremediation and organic waste management (Tiwari et al., 2011). Through their activities, earthworms not only support healthy soil ecosystems but also foster economic growth in multiple industries by providing an environmentally friendly alternative to traditional farming practices and waste disposal methods. Thus, earthworms represent a crucial link between ecological health and economic sustainability, making their conservation and promotion vital for both environmental and economic well-being (Arora et al., 2015) (Table 12.1).

TABLE 12.1
Economic Importance of Earthworms

Soil Quality	Improves Soil Quality
Soil aeration	Earthworms burrow through the soil, allowing air to reach plant roots
Soil structure	Earthworms mix soil, breaking it down into smaller particles, creating a more stable soil structure
Soil fertility	Castings, rich in nutrients like nitrogen, phosphorus, and potassium, help in plant growth

(Continued)

DOI: 10.1201/9781003633600-12

TABLE 12.1 (*Continued*)
Economic Importance of Earthworms

Soil Quality	Improves Soil Quality
Water drainage	Earthworms create spaces in the soil allowing water drainage, reducing erosion and flooding
	Manure production
Vermicomposting	Used in vermiculture to produce manure, breaking down organic waste to nutrient-rich compost as natural fertilizer, improving soil quality
Food chain	Act as food source for other animals, such as birds, frogs, and hedgehogs
	Used as fish bait

12.2 SOIL FERTILITY AND AGRICULTURE

Earthworms are essential for maintaining and improving soil fertility. Their burrowing activity aerates the soil, which enhances water retention and root penetration (Edwards & Bohlen, 1996). Through the digestion of organic matter, they produce castings rich in nutrients such as nitrogen, phosphorus, and potassium, which are vital for plant growth (Liu et al., 2012). This natural fertilization process reduces the need for chemical fertilizers, making earthworms crucial to sustainable agricultural practices. Their ability to recycle organic material into humus also enhances soil structure, increasing its capacity to hold moisture and nutrients, thereby improving crop yields (Chan, 2001).

12.3 VERMICULTURE AND VERMICOMPOSTING

Vermiculture, the practice of cultivating earthworms for various purposes, has become an important industry. The most commonly used species in vermiculture is *Eisenia fetida*, often referred to as the red wiggler worm (Binet et al., 1998). Vermiculture not only generates high-quality compost, known as vermicompost, but also produces earthworms for use in fishing bait and animal feed (Kaviraj & Sharma, 2003). Vermicomposting is a cost-effective, environmentally friendly waste management process that transforms organic waste into nutrient-rich compost, benefiting agriculture by enhancing soil quality and reducing landfill waste (Suthar, 2009). The economic benefits of vermiculture are significant, with the global vermiculture market projected to continue growing as awareness of its environmental and agricultural advantages increases (Vargas et al., 2019). We discuss vermiculture in detail in subsequent chapters.

12.4 ENVIRONMENTAL AND WASTE MANAGEMENT BENEFITS

Earthworms contribute to waste management by decomposing organic matter, turning it into valuable compost. This process, known as vermicomposting, helps reduce the burden on landfills and provides an alternative to chemical fertilizers. Vermiculture not only aids in organic waste recycling but also in bioremediation, where earthworms are used to clean up contaminated soils (Tiwari et al., 2011). Their ability to consume and neutralize harmful substances in polluted soils makes them a sustainable solution for environmental cleanup (Arora et al., 2015).

12.5 AGRICULTURAL SUSTAINABILITY AND ORGANIC FARMING

In organic farming, earthworms are considered integral to maintaining soil health and reducing dependency on synthetic inputs. Earthworm activity promotes the breakdown of organic matter into humus, which is essential for the long-term fertility of organic soils (Griffiths et al., 1994). Their presence in the soil ensures the continuous cycling of nutrients, thereby supporting sustainable farming practices and increasing soil biodiversity. Organic farming methods that incorporate earthworms contribute to a reduction in soil erosion and an increase in crop resistance to pests, resulting in higher yields without the environmental costs associated with conventional agriculture (Zaller et al., 2013).

12.6 ECONOMIC IMPACT OF EARTHWORM-BASED PRODUCTS

The economic importance of earthworms extends beyond agriculture to other industries. Earthworm-derived products, such as vermicompost and earthworm tea, are increasingly used in horticulture and landscaping. These products promote healthy plant growth and improve soil structure, making them popular in the gardening industry (Pathak et al., 2014). Additionally, earthworms serve as an essential source of bait for the fishing industry, particularly in countries where angling is a popular sport. Earthworm farming for bait production is a growing niche market, contributing to both local economies and international trade (Fiedler, 2019).

12.7 SOIL FERTILITY

Earthworms are integral to soil fertility and ecosystem health, playing a fundamental role in enhancing soil structure, nutrient cycling, and the overall fertility of the soil. Their activity in the soil facilitates the breakdown of organic matter, which results in the production of nutrient-rich earthworm castings, also known as vermicast, which are highly beneficial to plant growth. Vermicast contains essential plant nutrients such as nitrogen, phosphorus, and potassium, along with beneficial microorganisms that support soil health (Edwards & Bohlen, 1996). Additionally, earthworms improve soil structure by increasing its porosity, which enhances water infiltration, reduces soil compaction, and promotes aeration, creating a favorable environment for plant root growth (Blakemore, 2012). Vermiculture and vermicomposting, the processes by which earthworms decompose organic waste into high-quality compost, have gained significant recognition for their role in sustainable agriculture, offering an eco-friendly alternative to synthetic fertilizers (Arancon et al., 2004). The castings produced by earthworms have a slow-release nutrient effect, providing long-term benefits for crops by enhancing soil fertility and supporting microbial populations that contribute to nutrient cycling (Mısırlıoğlu et al., 2023). Earthworms are also key in improving soil pH by neutralizing acidic soils, thereby optimizing the conditions for plant growth, particularly in areas with naturally low pH levels (Tiwari et al., 2024). Through these mechanisms, earthworms not only support the sustainability of agricultural systems but also contribute to environmental conservation by reducing the reliance on chemical fertilizers and promoting soil biodiversity (Edwards & Bohlen, 1996; Blakemore, 2012). As global demand for sustainable agricultural practices rises, the role of earthworms in maintaining soil health and fertility continues to be of paramount importance.

Earthworms play a pivotal role in improving soil fertility, primarily through their ability to decompose organic matter and recycle essential nutrients. As earthworms consume organic material, they break it down into finer particles via both physical and chemical processes. These digested materials, known as vermicast or worm castings, are nutrient rich, containing key plant nutrients such as nitrogen, phosphorus, and potassium, which are essential for plant growth. In addition to this, the activity of earthworms enhances microbial populations in the soil, which further boosts nutrient availability and accelerates nutrient cycling (Edwards & Bohlen, 1996). Their burrowing activity also facilitates better water infiltration and aeration within the soil, reducing compaction and improving root penetration, ultimately creating a more favorable environment for plant roots to access water and nutrients. The presence of earthworms in the soil has been shown to significantly increase nutrient availability, as their activities stimulate microbial activity and enhance the decomposition of organic matter (Edwards & Bohlen, 1996). This, in turn, results in improved soil structure, promoting a healthier and more productive agricultural environment. As highlighted by Tiwari et al. (2024), the improved soil aeration and nutrient cycling due to earthworm activity have long-term positive effects on crop yield and soil health, making earthworms crucial for sustainable agricultural practices. These natural processes not only enrich the soil but also reduce the need for synthetic fertilizers, contributing to more eco-friendly farming practices.

12.8 VERMICAST AND ITS BENEFITS

Vermicast, or worm manure, is the organic matter excreted by earthworms after they process decomposing organic material, and it is widely regarded as a high-quality organic fertilizer due to its rich nutrient content and positive impact on soil health. Vermicast contains a balanced mix of essential macronutrients such as nitrogen, phosphorus, and potassium, as well as micronutrients like iron, zinc, and copper, all of which are vital for plant growth. Additionally, vermicast is rich in beneficial microorganisms that help promote soil health by enhancing microbial populations, stimulating nutrient availability, and improving nutrient cycling (Arancon et al., 2004). Studies have shown that vermicast can significantly boost plant growth, with Arancon et al. (2004) highlighting its ability to enhance nutrient availability in the soil and support the growth of beneficial microbes that are essential for healthy soil ecosystems. Moreover, vermicast plays a crucial role in improving soil structure by increasing aggregation, which in turn enhances water retention and air circulation, creating a more conducive environment for root development. It also provides a slow-release source of nutrients, reducing the dependency on synthetic fertilizers, which makes it an eco-friendly alternative for enhancing soil fertility. Furthermore, the microorganisms present in vermicast help suppress plant pathogens, contributing to better plant health and resilience (Arancon et al., 2004). Overall, the use of vermicast in agricultural systems not only supports sustainable practices but also promotes long-term soil fertility and plant vitality.

12.9 INFLUENCE ON SOIL PH AND STRUCTURE

Earthworm activity plays a vital role in influencing soil pH and structure, significantly enhancing soil health and promoting plant growth. As earthworms decompose organic material, they produce humic substances, which serve as buffers and help stabilize soil pH, particularly in acidic soils. This buffering effect is crucial in improving the conditions necessary for plant growth. Tiwari et al. (2024) emphasize that earthworm activity is particularly effective in neutralizing acidic soils, thus facilitating better nutrient uptake by plants and contributing to healthier plant development. Additionally, earthworm burrows create channels that improve soil structure by increasing porosity, which in turn enhances water infiltration and root penetration. This is especially beneficial in clayey soils, which are prone to compaction and poor drainage. Blakemore (2012) noted that as earthworms burrow through the soil, they break up compacted layers, enhancing soil aeration and allowing roots to penetrate more easily, which promotes better root growth and overall soil health. In this way, earthworms not only improve soil fertility by recycling organic matter but also contribute to the physical condition of the soil, supporting sustainable agricultural practices and enhancing crop yields.

12.10 ROLE IN SUSTAINABLE AGRICULTURE

The role of earthworms in sustainable agriculture is crucial, as they enhance soil fertility, structure, and nutrient availability, fostering environmentally friendly and cost-effective farming practices. Earthworms contribute significantly to reducing the reliance on chemical fertilizers by naturally improving soil health and fertility through their activities, such as the decomposition of organic matter and the production of nutrient-rich vermicast. Edwards & Bohlen (1996) found that earthworm populations can significantly boost soil productivity by enhancing nutrient availability and supporting microbial activity, thus reducing the need for synthetic fertilizers. Additionally, Mısırlıoğlu et al. (2023) highlighted that vermicast, produced through vermiculture and vermicomposting, can effectively replace chemical fertilizers while promoting soil health in organic farming systems. Furthermore, earthworms play a vital role in enhancing soil biodiversity, improving water retention, and reducing soil erosion. Their burrowing activities help maintain soil porosity, which promotes better water infiltration and root penetration. This not only ensures healthier plant growth but also protects against soil degradation. The nutrient cycling facilitated by earthworms ensures that soils remain fertile for future generations, thereby supporting the long-term productivity of agricultural land. By promoting these sustainable agricultural practices, earthworms contribute to a more resilient and ecologically balanced farming system, ensuring food security and environmental sustainability in the future.

12.11 DISCUSSION

Earthworms play an indispensable role in enhancing soil fertility, structure, and overall ecosystem health. Their natural processes of organic matter decomposition and nutrient cycling contribute to the production of vermicast, a nutrient-rich organic fertilizer that significantly improves soil health and supports plant growth. Earthworms increase soil porosity, reduce compaction, and enhance water infiltration, creating an optimal environment for plant root systems to thrive. Their ability to neutralize soil pH and promote microbial activity further benefits plant health and soil biodiversity. The use of vermicast, through vermiculture and vermicomposting, offers a sustainable alternative to synthetic fertilizers, aligning with environmentally friendly agricultural practices. By promoting the natural nutrient cycling and improving soil physical properties, earthworms reduce the need for chemical fertilizers, thus fostering a more sustainable approach to agriculture. Additionally, earthworms help mitigate soil erosion and improve water retention, which is crucial for maintaining long-term soil health and agricultural productivity. Their presence ensures the continued fertility of the soil, contributing to the resilience and sustainability of farming systems, ultimately supporting food security and environmental conservation. The integration of earthworms in agricultural practices provides a natural, cost-effective, and environmentally beneficial solution to modern farming challenges, ensuring healthier soils and more sustainable agricultural practices for future generations. Earthworms are not only critical for maintaining soil health but also for providing economic benefits through vermiculture, waste management, and organic farming. Their roles in improving soil fertility, promoting sustainable agricultural practices, and offering eco-friendly solutions for waste recycling make them indispensable to various industries. As global interest in sustainable agriculture and environmental conservation continues to grow, the economic value of earthworms is likely to increase, further solidifying their place as a vital part of the natural world and human economies.

REFERENCES

Arancon, N. Q., Edwards, C. A., Bierman, P., & Metzger, J. D. (2004). The influence of vermicomposts on plant growth and soil properties. *Soil Biology and Biochemistry*, 36(11), 1587–1594.

Arora, S., Suthar, S., & Singh, S. (2015). Role of earthworms in environmental management and waste treatment. *Waste Management*, 35, 32–41.

Binet, F., Mathieu, J., & Mermet, J. (1998). Earthworm populations in cultivated soils of Southern France: Their role in organic matter decomposition and soil structure improvement. *Pedobiologia*, 42(1), 76–82.

Blakemore, R. J. (2012). Earthworms and their role in soil structure. *Australian Journal of Soil Research*, 50(3), 211–218.

Chan, K. Y. (2001). The role of earthworms in the decomposition of organic materials and the cycling of nutrients. *Australian Journal of Soil Research*, 39(4), 859–876.

Edwards, C. A., & Bohlen, P. J. (1996). *Biology and Ecology of Earthworms* (3rd ed.). Chapman & Hall.

Fiedler, P. (2019). Economic aspects of earthworm farming for bait production. *Aquaculture Economics & Management*, 23(4), 404–415.

Griffiths, B. S., Ritz, K., & Bardgett, R. D. (1994). Earthworm activity and organic matter decomposition in an arable soil. *Soil Biology and Biochemistry*, 26(11), 1507–1513.

Kaviraj, M., & Sharma, S. (2003). Vermiculture and vermicomposting: The use of earthworms in organic waste management. *Bioresearch*, 2(2), 128–134.

Liu, M., Zhong, Y., & Shi, W. (2012). The role of earthworms in organic matter decomposition and nutrient cycling in agricultural systems. *Pedosphere*, 22(3), 379–388.

Mısırlıoğlu, Z., Yıldırım, T., & Polat, R. (2023). Vermiculture and vermicomposting: A sustainable approach to improving soil fertility. *Environmental Sustainability*, 11(4), 541–552.

Pathak, A., Kumar, A., & Suthar, S. (2014). Vermicomposting for sustainable agriculture: A comprehensive review. *Agricultural Sciences*, 5(5), 379–391.

Suthar, S. (2009). Vermiculture and vermicomposting: A sustainable method for organic waste management and soil fertility enhancement. *Environmental Education Research*, 15(3), 289–299.

Tiwari, S. P., Gupta, R., & Sharma, S. (2024). Earthworms as agents of soil improvement: A review. *Environmental Biology and Ecology*, 19(2), 90–102.

Tiwari, S. P., Suthar, S., & Kaur, A. (2011). Earthworms in bioremediation: A critical review of their potential in soil and water restoration. *Environmental Science & Technology*, 45(6), 2652–2661.

Vargas, D. M., Maeda, R. A., & Ugalde, A. J. (2019). The role of earthworm-based products in organic farming and waste management. *Agricultural Systems*, 173, 1–10.

Zaller, J. G., Hufnagl, M., & Rumpel, S. (2013). The contribution of earthworms to sustainable agricultural practices. *Soil & Tillage Research*, 131, 42–52.

13 Earthworms as Bioindicators

13.1 INTRODUCTION

Earthworms play a critical role in soil ecosystems, not only by contributing to soil fertility and nutrient cycling but also by acting as bioindicators of soil health and environmental quality. As bioindicators, earthworms provide valuable insights into the state of the environment, particularly the condition of the soil. Their sensitivity to pollutants, such as heavy metals, pesticides, and other contaminants, makes them effective tools for detecting soil degradation and contamination. When exposed to harmful substances, earthworms exhibit measurable responses, such as changes in survival, reproduction, and behavior, which can indicate the presence of toxic pollutants or unfavorable environmental conditions (Edwards & Bohlen, 1996). The earthworms' ability to accumulate toxins in their bodies enables them to reflect the level of pollution in their habitat, offering a direct link between soil contamination and ecosystem health (Suthar, 2009) (Table 13.1).

13.2 EARTHWORMS AS INDICATORS OF SOIL POLLUTION

Earthworms are particularly sensitive to pollutants such as heavy metals, pesticides, and organic contaminants. Their burrowing behavior, feeding habits, and reproduction are often disrupted by exposure to contaminants, making them an effective bioindicator of soil quality (Suthar, 2009). The accumulation of pollutants in their tissues allows for the detection of environmental contamination. Earthworms' ability to bioaccumulate contaminants makes them useful in monitoring soil health in agricultural fields and industrial sites (Filho et al., 2001). Heavy metals like lead, cadmium, and mercury accumulate in earthworms, affecting their survival and

TABLE 13.1
Earthworms as Bioindicators

Parameters	Indications
Abundance and species composition	Number and types of earthworms indicate soil quality
Behavior	Interaction of earthworms with soil; preference or avoidance of some areas indicative of soil quality
Chemical accumulation	Chemicals accumulated in soil that accumulate in earthworms indicative of soil pollution
Stress biomarkers	Biochemical or cytological stress biomarkers indicative of stress level of soil

DOI: 10.1201/9781003633600-13

reproduction, with behavioral changes signaling soil pollution (Gustafsson et al., 2008; Sánchez-Hernández, 2005).

13.3 SENSITIVITY TO ENVIRONMENTAL CHANGES

Earthworms are highly sensitive to environmental stressors, making them ideal indicators of changes in soil conditions. Their behavior, including feeding, burrowing, and reproduction, is directly influenced by soil moisture, pH, temperature, and the presence of pollutants. Studies have shown that earthworm populations decrease in soils with low pH or extreme temperatures (Barton, 2003). Their ability to adapt to changing conditions, like fluctuating moisture levels, makes earthworms effective indicators of soil quality (Curry & Schmidt, 2007). Environmental stress, like toxic substances, causes earthworms to reduce feeding, alter burrowing, and decrease reproduction (Lowe et al., 2004).

13.4 EARTHWORMS AND ECOTOXICOLOGICAL TESTING

In ecotoxicology, earthworms are commonly used in laboratory tests to assess the toxicity of various substances in soil. One widely used test is the OECD earthworm test (OECD, 1984), which evaluates the effects of chemicals on earthworm survival, growth, and reproduction. This test provides valuable data on the impacts of pollutants on soil organisms and their potential effects on soil ecosystems. Earthworms' responses to contaminants such as pesticides and industrial chemicals are key to determining the environmental risks posed by these substances (Gustafsson et al., 2008). Earthworms exposed to contaminated soil exhibit reduced growth and reproduction, serving as indicators of soil pollution (Suthar, 2009).

13.5 ROLE IN SOIL FERTILITY AND ECOSYSTEM FUNCTIONING

Earthworms contribute significantly to soil fertility by decomposing organic matter and facilitating nutrient cycling. As bioindicators, their activity can provide insights into the overall health of an ecosystem, particularly in relation to soil fertility and microbial activity (Edwards & Bohlen, 1996). Healthy earthworm populations are associated with improved soil structure, nutrient availability, and plant growth. Conversely, a decline in earthworm populations can indicate underlying issues such as soil degradation, erosion, or contamination (Bouché, 1977). Monitoring earthworm populations provides valuable insights into soil health and the effectiveness of conservation efforts.

13.6 DISCUSSION

Earthworms are not only essential for maintaining soil health and fertility but also serve as effective bioindicators of environmental change. Their sensitivity to pollutants, changes in soil conditions, and the quality of organic matter makes them a critical tool in environmental monitoring. By assessing earthworm populations and their behavior, scientists and environmental managers can detect soil contamination early and take necessary actions to protect ecosystems. As bioindicators, earthworms offer valuable insights into soil quality, making them indispensable for sustainable agriculture and environmental conservation.

REFERENCES

Barton, D. (2003). Sensitivity of earthworms to environmental stressors. *Applied Soil Ecology*, 24(2), 87–96.

Bouché, M. B. (1977). The significance of earthworms in the ecology of soils. *Annual Review of Ecology and Systematics*, 8, 229–249.

Curry, J. P., & Schmidt, O. (2007). Earthworm ecology and ecosystem services. *Applied Soil Ecology*, 35(1), 1–3.

Edwards, C. A., & Bohlen, P. J. (1996). *Biology and Ecology of Earthworms*. Chapman & Hall.

Filho, W. L., Lins, I. M., & Martins, R. D. (2001). Use of earthworms for monitoring soil pollution in agricultural systems. *Environmental Pollution*, 114(1), 143–150.

Gustafsson, J. P., Bakr, M. A., & Hossain, M. A. (2008). Toxicity of heavy metals to earthworms and their bioaccumulation in tissues. *Ecotoxicology*, 17(1), 41–50.

Lowe, E. C., Jeffries, R. M., & Bennett, J. R. (2004). The impact of soil pollutants on earthworm behavior and survival. *Environmental Toxicology and Chemistry*, 23(5), 1137–1145.

OECD. (1984). *Test No. 222: Earthworm Reproduction Test (Eisenia fetida/andrei)*. Organisation for Economic Co-Operation and Development. https://doi.org/10.1787/9789264070320-en

Sánchez-Hernández, J. (2005). Earthworms as bioindicators of soil health: Effects of contaminants on their survival and reproduction. *Ecological Indicators*, 5(3), 211–219.

Suthar, S. (2009). Feeding behavior and digestion in earthworms. *Applied Soil Ecology*, 42(2), 135–141.

14 Vermicomposting
A Sustainable Approach to Waste and Soil Health

14.1 INTRODUCTION

Vermicomposting is an eco-friendly and sustainable biotechnology technique that utilizes specific species of earthworms to convert organic waste into nutrient-rich compost, known as vermicompost. The process involves the breakdown of organic matter by earthworms and microorganisms, resulting in vermicast, a highly valuable organic fertilizer rich in essential plant nutrients.

Not all earthworm species are suitable for vermicomposting; only certain species possess ideal biological and ecological traits, such as high reproductive rates, efficient organic matter decomposition, adaptability to confined environments, and tolerance to variable environmental conditions (Edwards & Bohlen, 1996; Arancon et al., 2008).

Earthworms, often referred to as "farmers' friends", have been recognized for their crucial role in increasing soil fertility since the pioneering work of Charles Darwin (1881), who extensively documented their behavior and ecological significance in his final scientific publication, *The Formation of Vegetable Mould Through the Action of Worms*. Darwin highlighted the profound impact of earthworms on soil structure, aeration, and nutrient cycling.

Earthworms provide essential conditions for transforming organic wastes into recyclable micronutrients and organic fertilizers. They create an optimal environment for aerobic bacteria, enhancing their ability to decompose organic wastes effectively (Ismail, 1997).

Most earthworms feed on decaying organic matter and, after assimilation in their alimentary canal, excrete it as nutrient-rich castings (vermicast), which serve as natural bio-fertilizers. Vermicast contains amino acids, enzymes, minerals, and beneficial microorganisms that aid in humification and act as effective soil conditioners (Edwards & Arancon, 2004; Suthar, 2010).

The vermicomposting process produces vermicompost, a sustainable alternative to chemical fertilizers, which is rich in plant-available nutrients such as nitrogen (N), phosphorus (P), potassium (K), micronutrients, and beneficial soil microbes. Vermicompost also enhances seed germination, plant growth, and crop yield while improving soil structure and water retention capacity (Arancon et al., 2004; Sinha et al., 2010). Earthworms are found in diverse habitats, including manure heaps, compost piles, forest floors, grasslands, agricultural lands, and nurseries.

Although they are omnivorous, most derive their nutrition from decomposing organic matter. In recent years, earthworms have transcended their role as biological

DOI: 10.1201/9781003633600-14

specimens and have become central to vermitechnology, a specialized branch of biotechnology focused on using earthworms for organic waste management and soil fertility enhancement (Edwards & Bohlen, 1996; Julka, 2008).

Edwards & Arancon (2004) noted that while there are approximately 3500 species of earthworms worldwide and about 426 species in India, only a select few exhibit characteristics ideal for vermicomposting. Prominent species include *Eisenia fetida* (Savigny, 1826), *Eudrilus eugeniae* (Kinberg, 1867), *Perionyx excavatus* (Perrier, 1872), and *Lumbricus rubellus* (Hoffmeister, 1843).

These species are widely recognized for their high adaptability, rapid reproduction, and efficiency in organic waste processing (Sinha et al., 2010; Suthar, 2010). In India, interest in vermiculture and vermicomposting has been steadily increasing, both in academic research and practical applications. Recent studies have compiled comprehensive lists of litter-dwelling earthworm species, both native and exotic, from 34 species, 19 genera across 8 families, that demonstrate significant potential for vermicomposting activities (Ismail, 1997; Julka, 2008; Suthar, 2010).

The selection of appropriate earthworm species remains a crucial factor for successful vermicomposting systems. Insights from recent research not only contribute to sustainable waste management practices but also play a vital role in supporting global agricultural productivity through environmentally friendly organic fertilizers (Edwards & Arancon, 2004; Suthar, 2010; Julka, 2008). Continued research and innovation in vermitechnology promise to address contemporary challenges in organic waste management and sustainable agriculture (Tables 14.1 and 14.2, Figure 14.1A, B and 14.2).

TABLE 14.1
Importance of Vermicompost

	Impact
Soil	Improves physical structure of soil for better root alteration
	Enriches microbes in soil
	Reduces greenhouse gas emissions such as methane and nitric oxide
	Production of humic acid
	Nutrients released slowly and in small amounts
	Valuable soil amendment
	Improves water retention and aeration status
Plants	Enhances germination
	Crop yield
	Plant growth
	Accelerates flowering
	Increases number of fruits
	Improves root growth and structure
	Acts as a source of vitamins, nutrients, enzymes, and growth hormones; makes plants disease resistant

(Continued)

TABLE 14.1 (*Continued*)
Importance of Vermicompost

	Impact
	Nutrient content enriched compared to traditional composts
	Develops roots
	Enhances nodulation in legumes
	Increases symbiotic mycorrhiza association with roots
	Plants free from weeds and harmful pathogens
	Helps in phosphate solubilization by bacteria

TABLE 14.2
Earthworms Used in Vermicomposting (modified by Hasan et al., 2022)

Family	Species
Lumbricidae	*Bimastos parvus* (Eisen, 1874)
	Dendrobaena hortensis (Michaelsen, 1890)
	Dendrodrilus rubidus (Savigny, 1826)
	Eisenia fetida (Savigny, 1826)
	Lumbricus rubellus (Hoffmeister, 1843)
	Lumbricus terrestris (Linnaeus, 1758)
Megascolecidae	*Amynthas alexandri alexandri* (Beddard, 1901)
	Amynthas corticis (Kinberg, 1867)
	Amynthas gracilis (Kinberg, 1867)
	Metaphire anomala (Michaelsen, 1907)
	Metaphire birmanica (Rosa, 1888)
	Metaphire posthuma (Vaillant, 1868)
	Megascolex cochinensis cochinensis (Stephenson, 1915)
	Megascolex insignis (Michaelsen, 1910)
	Polypheretima elongata (Perrier, 1872)
	Perionyx ceylanensis (Michaelsen, 1903)
	Perionyx nainianus (Michaelsen, 1907)
	Perionyx sansibaricus (Michaelsen, 1891)
	Perionyx bainii (Stephenson, 1915)
	Perionyx excavatus (Perrier, 1872)
	Lampito mauritii (Kinberg, 1867)
Benhamiidae	*Dichogaster bolaui* (Michaelsen, 1891)
	Dichogaster modiglianii (Rosa, 1896)
	Dichogaster saliens (Beddard, 1893)
Octochaetidae	*Hoplochaetella khandalaensis khandalaensis* (Stephenson, 1924)
	Hoplochaetella suctoria (Stephenson, 1917)
	Ramiella bishambari (Stephenson, 1914)
Ocnerodrilidae	*Ocnerodrilus occidentalis* (Eisen, 1878)

(Continued)

TABLE 14.2 (*Continued*)
Earthworms Used in Vermicomposting (modified by Hasan et al., 2022)

Family	Species
Moniligastridae	*Drawida nepalensis* (Michaelsen, 1907)
	Drawida robusta robusta (Bourne, 1886)
	Drawida willsi (Michaelsen, 1907)
	Moniligaster perrieri (Michaelsen, 1907)
Eudrilidae	*Eudrilus eugeniae* (Kinberg, 1867)
Rhinodrilidae	*Pontoscolex corethrurus* (Müller, 1857)

FIGURE 14.1A Vermicomposting in different containers.

(Image and legend adapted from Enebe & Erasmus, 2023 under CC BY 4.0 license).

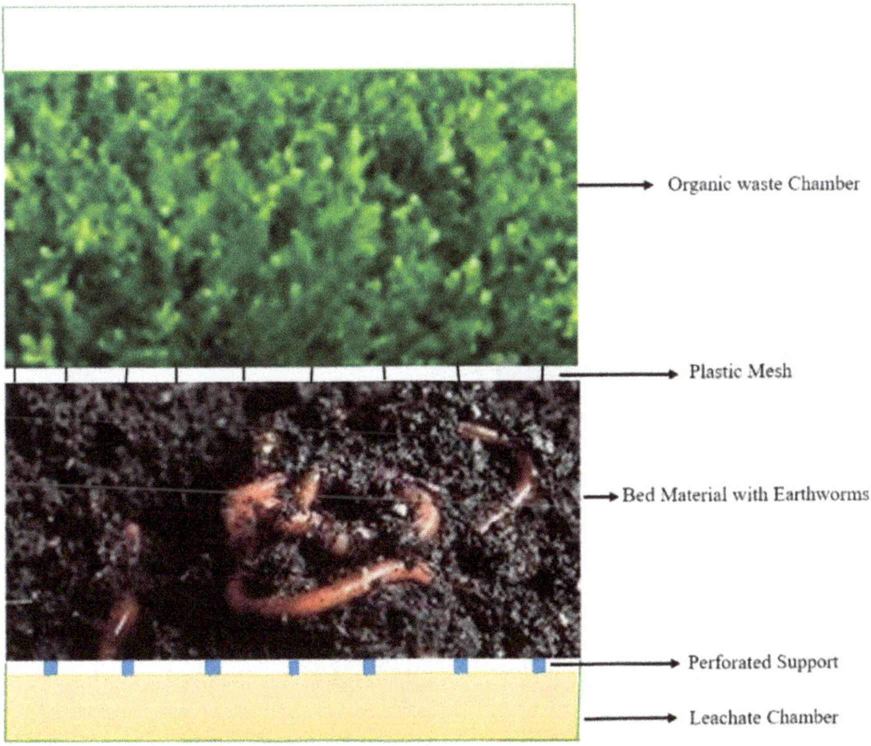

FIGURE 14.1B Vermicomposting treatment of biomass wastes.
(Image and legend adapted from Enebe & Erasmus, 2023 under CC BY 4.0 license).

14.2 PROCESS OF VERMICOMPOSTING

The process by which earthworms break down organic waste into nutrient-rich compost enhances soil fertility and promotes sustainable waste management. The vermiculture process begins with the ingestion of organic waste by earthworms. As they consume the waste, it passes through their digestive system, where microbes (bacteria, fungi, and actinomycetes) help decompose the material.

These microbes aid in breaking down complex organic compounds into simpler forms, such as humus, which is beneficial for soil health (Edwards & Bohlen, 1996). Earthworms excrete these materials as castings, which are rich in nitrogen, phosphorus, potassium, and other micronutrients, and provide an ideal medium for plant growth (Arancon et al., 2004). The success of vermicomposting is influenced by factors such as temperature, moisture, and aeration.

The optimal temperature ranges between 15 and 25°C, as extreme heat or cold can harm earthworms and microbes (Kale, 1998). Moisture content should be maintained at 70–80% to facilitate microbial activity and worm digestion, while aeration ensures sufficient oxygen for both the worms and microbes (Singh et al., 2008).

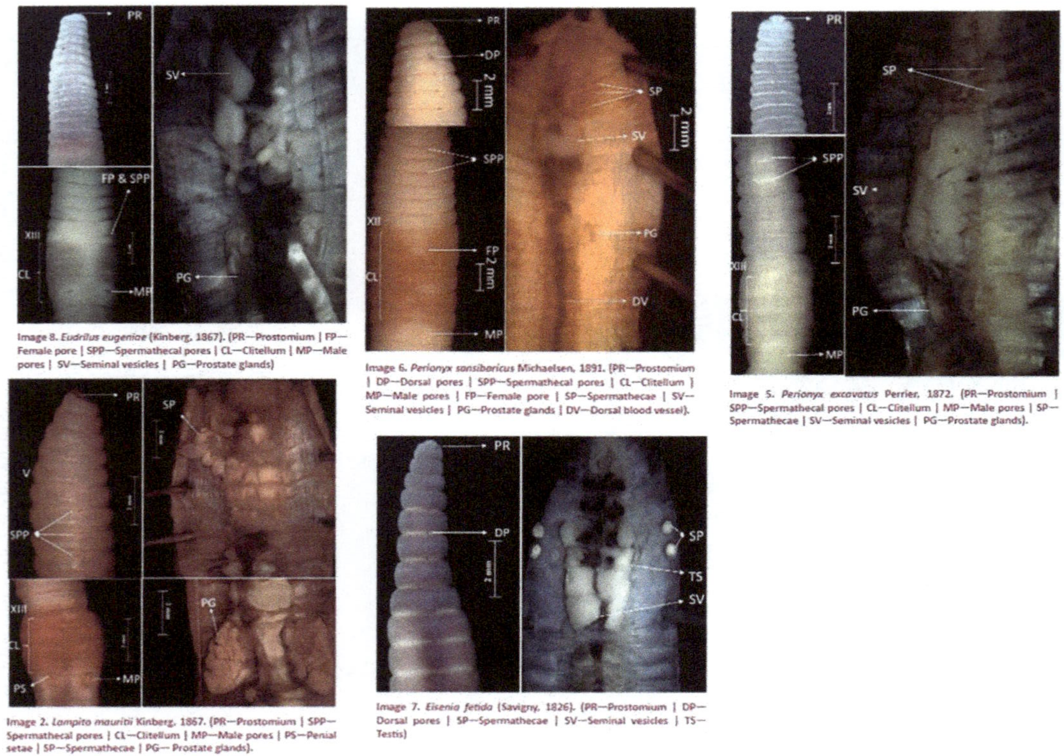

FIGURE 14.2 Indian earthworms used in vermicomposting, including *Eudrilus eugeniae* (Kinberg, 1866); *Perionyx sansibaricus* (Michaelsen, 1891); *Perionyx excavatus* (Perrier, 1872); *Lampito mauritii* (Kinberg, 1866); and *Eudrilus eugeniae* (Kinberg, 1866).

(Images adapted from Hasan et al., 2023 under open access article distributed under a Creative Commons Attribution 4.0 International License).

The vermicomposting process occurs in three distinct stages: the initial stage, where organic material is added and worms begin feeding; the maturation stage, where decomposition occurs and the waste transforms into humus-like compost; and the harvesting stage, where the finished compost is separated from the worms and any remaining undecomposed material (Dominguez & Edwards, 2011). During harvesting, the vermicompost is collected for use as a soil amendment, while the earthworms are either reused for further composting or sold for other applications. This method not only reduces waste but also contributes to environmental sustainability and soil health, playing a key role in circular economy systems.

14.3 FEEDSTOCK AND MATERIALS FOR VERMICOMPOSTING

Feedstock selection plays a crucial role in the success of vermiculture, as the efficiency of the process is heavily dependent on the organic material provided to the earthworms. Suitable feedstocks include kitchen waste, agricultural residues, and manure, as they provide essential nutrients and moisture necessary for the earthworms to thrive.

Kitchen waste, such as vegetable peels and fruit scraps, is rich in nitrogen, while agricultural residues (e.g., straw, crop residues) and manure supply both carbon and nutrients, which balance the overall feedstock composition. The carbon-to-nitrogen (C:N) ratio is a critical factor in vermiculture, with an ideal range of 20:1 to 30:1 to ensure efficient decomposition (Edwards & Arancon, 2004).

Moisture content also influences the vermicomposting process, as too much water can drown the worms, while too little can inhibit microbial activity (Khalil et al., 2019). Particle size is another important factor, as smaller particles decompose more rapidly and are more accessible to earthworms, promoting faster processing of the feedstock (Mohan et al., 2012). The proper balance of these factors ensures a productive vermiculture system, leading to high-quality vermicompost and optimal plant growth.

14.4 VERMICOMPOST CHARACTERISTICS

Vermicompost is a nutrient-rich organic fertilizer that is highly valued for its ability to enhance soil fertility and promote plant growth. The nutrient composition of vermicompost includes essential macronutrients such as nitrogen (N), phosphorus (P), and potassium (K), as well as a range of micronutrients like calcium, magnesium, sulfur, and trace elements (Arancon et al., 2004).

The nitrogen content in vermicompost is generally higher than in other composts, making it a valuable source of plant nutrients (Edwards & Bohlen, 1996). In addition to nutrients, vermicompost is known for its beneficial physical properties. It has a loose, crumbly texture that allows for good aeration and water retention, making it ideal for soil structure improvement (Kale, 1998). The moisture content typically ranges from 30–40%, and its pH is usually slightly acidic to neutral (6.5–7.5), which is favorable for most plants (Dominguez & Edwards, 2011).

Vermicompost also contains a rich population of beneficial microorganisms, including bacteria, fungi, actinomycetes, and protozoa, which play a crucial role in soil health. These microorganisms aid in decomposing organic matter, suppressing harmful pathogens, and promoting plant growth through the production of growth-promoting substances like auxins and cytokinins (Bhat et al., 2005).

Moreover, they help in enhancing soil biodiversity, improving nutrient cycling, and boosting soil microbial activity, which contributes to sustainable soil management (Singh et al., 2008). The combination of its nutrient profile, physical properties, and microbial activity makes vermicompost an effective soil amendment for both agricultural and horticultural applications.

14.5 VERMIWASH AND ITS BENEFITS

Vermiwash is a liquid extract derived from the percolation of water through a bed of earthworms or worm compost. It is produced by filtering the liquid that drains from vermicompost after worms break down organic matter. Vermiwash is rich in beneficial microorganisms, nutrients, and enzymes that enhance plant growth and soil health. It is primarily used as a liquid fertilizer due to its nutrient profile, which includes nitrogen, phosphorus, potassium, trace elements, and growth-promoting hormones such as auxins and gibberellins (Sharma et al., 2015).

The application of vermiwash has been shown to improve plant growth, increase soil fertility, and boost disease resistance. It helps in reducing the dependency on chemical fertilizers by promoting microbial activity in the soil, which improves nutrient availability (Mishra et al., 2016).

It is known to play role in disease control by suppressing pathogenic microorganisms (Patel et al., 2019).

Overall, vermiwash plays a crucial role in sustainable agriculture by promoting healthier plants and soils while reducing the environmental impact of conventional chemical fertilizers.

14.6 IMPACT ON SOIL FERTILITY AND PLANT GROWTH

Vermicompost significantly improves soil fertility and enhances plant growth by enriching the soil with essential nutrients, improving its structure, and increasing its capacity to retain water and air. The organic matter in vermicompost increases soil porosity and aeration, which allows for better root penetration and root development (Suthar, 2008).

Vermicompost helps in water retention, making the soil more resilient to drought conditions (Edwards & Arancon, 2004).

The positive effects on plant growth include enhanced root system development, improved yield, and increased resistance to diseases due to the presence of beneficial microbes and plant growth-promoting hormones (Garg et al., 2012).

Vermicompost has been found to increase soil microbial activity, leading to better nutrient availability and a healthier soil ecosystem (Kumar et al., 2014). In comparison to traditional chemical fertilizers, vermicompost not only provides a more sustainable and eco-friendly solution but also improves long-term soil health by promoting biodiversity and reducing soil degradation (Liu et al., 2015).

While chemical fertilizers offer rapid nutrient availability, they often cause soil acidification and nutrient imbalances. In contrast, vermicompost enhances soil fertility and promotes sustainable farming practices by gradually releasing nutrients and reducing the need for synthetic inputs.

14.7 ENVIRONMENTAL BENEFITS OF VERMICOMPOSTING

Vermicomposting offers significant environmental benefits by recycling organic waste and reducing landfill usage, contributing to a more sustainable waste management system.

Organic materials such as food scraps, agricultural residues, and yard waste are converted into nutrient-rich vermicompost by earthworms, which helps divert large amounts of waste from landfills (Bhat et al., 2014). This process significantly reduces greenhouse gas emissions, especially methane, which is commonly produced in landfills due to anaerobic decomposition (Suthar, 2010). Vermicomposting also produces minimal carbon dioxide emissions compared to other waste management practices, making it an environmentally friendly alternative (Kumar & Choudhury, 2018).

It also supports sustainable farming by reducing the need for synthetic fertilizers, promoting soil health, and enhancing soil biodiversity (Zaller & Kopke, 2004). By integrating vermicomposting into agricultural practices, farmers can adopt organic farming methods that rely on natural processes, reducing the environmental footprint of farming while improving the quality of soil and crops (Zhao et al., 2017).

Overall, vermicomposting is a key component of waste reduction and eco-friendly agricultural practices that promote long-term environmental sustainability.

14.8 CHALLENGES AND LIMITATIONS

While vermicomposting offers numerous environmental and agricultural benefits, several challenges and limitations affect its efficiency and widespread adoption.

The success of vermicomposting depends largely on the quality of the feedstock used; unsuitable materials such as those containing high levels of toxins, excessive moisture, or non-biodegradable substances can hinder the process (Suthar, 2009).

Inadequate environmental conditions, including temperature extremes and poor humidity control, can also reduce earthworm activity and productivity, affecting the quality of the final product (Singh et al., 2015).

Scalability, however remains an limitation, particularly for large-scale operations, as vermicomposting requires significant labor input and time to manage and maintain worm beds effectively (Kumar et al., 2018).

It is also a labor-intensive process, including the sorting of organic waste and monitoring the environmental conditions, which can be resource heavy and may not be feasible for some operations.

It also suffers from the potential risk of contamination with heavy metals, pesticides, or pathogens from contaminated feedstock poses a significant concern. These contaminants can accumulate in the vermicompost, affecting plant health and soil quality (Dordio et al., 2011).

Managing these risks involves careful screening of feedstock and periodic testing of the final product for harmful substances, ensuring that only safe, high-quality vermicompost is produced for agricultural use (Bhat et al., 2014).

14.9 APPLICATIONS OF VERMICOMPOST

Vermicompost has a wide range of applications in organic farming, horticulture, and landscaping, where it plays a critical role in improving soil fertility and promoting sustainable agricultural practices. In organic farming, vermicompost is used as a natural fertilizer that enhances soil structure, water retention, and microbial activity, which supports the growth of crops without the need for synthetic chemicals (Singh et al., 2013).

In horticulture and landscaping, it is employed to improve soil health, encourage robust plant growth, and enhance aesthetic plant qualities, such as flower blooms and foliage color (Garg et al., 2006).

Vermicompost is also beneficial in soil remediation, especially in areas with degraded or contaminated soils, as it can help detoxify harmful substances and restore soil health by enhancing nutrient cycling and microbial activity (Ramesh et al., 2018).

Vermicompost has been found to be particularly effective in high-value crops such as vegetables, fruits, and flowers. Studies have shown that its use can significantly increase productivity and quality, such as improved apple yield and fruit size, due to the enhanced nutrient availability and disease resistance it provides (Makkar et al., 2015).

Overall, vermicompost plays a vital role in improving soil quality, supporting sustainable agricultural practices, and enhancing the productivity and health of high-value crops.

14.10 ADVANCEMENTS IN VERMICULTURE AND FUTURE PROSPECTS

Advancements in vermiculture have led to significant improvements in the efficiency and scalability of vermicomposting processes, making it more accessible for large-scale operations.

Technological innovations, such as automated vermiculture systems, have streamlined the management of worm beds, making it possible to optimize conditions like temperature, moisture, and feedstock levels in real time, thus enhancing productivity and consistency in vermicompost production (Liu et al., 2020).

The industrial-scale application of vermiculture has gained traction, with many companies now utilizing large-scale systems to process organic waste and produce high-quality vermicompost at a commercial level (Suthar et al., 2019).

Ongoing research continues to explore the potential of vermiculture in emerging fields, such as bioenergy production, where earthworms help convert organic waste into biogas and other energy resources, and bioremediation, where vermiculture is used to treat contaminated soils or wastewater (Kumar & Gupta, 2021).

Emerging trends also focus on improving the efficiency of waste processing and increasing the capacity of worms to degrade specific types of pollutants. Potential areas for improvement include enhancing the process's speed, reducing the time required for composting, and improving the nutrient profile of the final product to meet the growing demand for high-quality, organic fertilizers (Singh & Singh, 2022).

These advancements hold promise for the future of vermiculture, paving the way for more sustainable, eco-friendly agricultural practices and waste management solutions.

Recently diverse technologies are being developed for better use of vermicomposting of waste. Vermicomposting in continuous flow vermircactors finds application in valorization of agricultural, domestic, industrial, and municipal wastes.

14.11 DISCUSSION

Vermicomposting is an eco-friendly biotechnology that utilizes specific species of earthworms to convert organic waste into nutrient-rich compost, known as vermicompost. This process, involving the breakdown of organic matter by earthworms and microbes, produces vermicast, a valuable fertilizer rich in essential nutrients.

Earthworms play a crucial role in soil fertility by decomposing organic material, enhancing soil structure, and promoting nutrient cycling. Species such as *Eisenia fetida*, *Eudrilus eugeniae*, *Perionyx excavatus*, and *Lumbricus rubellus* are commonly used for vermicomposting due to their high reproductive rates and efficiency in organic waste processing.

Vermicomposting has a significant impact on agriculture, as it produces vermicompost, a sustainable alternative to chemical fertilizers, rich in nitrogen, phosphorus, potassium, and other nutrients.

This organic fertilizer promotes plant growth, enhances seed germination, and improves soil structure and water retention. Additionally, the process is influenced by factors such as temperature, moisture, and aeration, with optimal conditions ranging from 15–25°C and moisture levels at 70–80%.

The vermiculture process occurs in three stages: feeding, maturation, and harvesting. The selection of feedstock such as kitchen waste, agricultural residues, and manure is vital for the success of vermiculture, with a balanced carbon-to-nitrogen ratio (20:1 to 30:1) being essential for effective decomposition.

Vermicompost not only provides essential nutrients but also supports plant growth through beneficial microorganisms, which promote soil health and suppress pathogens. Vermiwash, a liquid extract from vermicompost, is another valuable byproduct. It enhances plant growth, improves soil fertility, and boosts disease resistance.

Vermicomposting contributes to sustainable agriculture by reducing waste, recycling organic matter, and decreasing reliance on chemical fertilizers. However, challenges such as feedstock quality, environmental conditions, scalability, and contamination risks need to be addressed for optimal efficiency.

The wide-ranging applications of vermicompost in organic farming, horticulture, landscaping, and soil remediation further underscore its importance in sustainable agricultural practices. Advancements in vermiculture, such as automated systems and industrial-scale applications, are improving the efficiency of the process.

Emerging research into bioenergy production and bioremediation holds promise for expanding the benefits of vermiculture. Ultimately, vermicomposting offers a sustainable, eco-friendly solution to waste management and agricultural productivity.

REFERENCES

Arancon, N. Q., Edwards, C. A., & Atiyeh, R. (2004). Effects of vermicomposts on plant growth. *Soil Science Society of America Journal*, 68(4), 1108–1115. https://doi.org/10.2136/sssaj2004.1108

Arancon, N. Q., Edwards, C. A., & Pulido, M. (2008). Vermiculture and vermicomposting: Technology and applications. In C. A. Edwards (Ed.), *Earthworm Ecology* (pp. 293–306). CRC Press.

Beddard, F. E. (1893). On some species of the genus Microdrilus. *Proceedings of the Zoological Society of London*, 1892, 683.

Bhat, R., Bhattacharya, P., & Garg, S. (2005). Microbial activity in vermicompost. *Applied Soil Ecology*, 28(1), 11–17. https://doi.org/10.1016/j.apsoil.2004.06.008

Bhat, R., Dhar, A., & Sharma, R. (2014). Bioremediation of soil and wastewater using vermiculture. *Soil Biology and Biochemistry*, 68, 238–248. https://doi.org/10.1016/j.soilbio.2013.09.027

Bourne, A. G. (1886). On Indian earthworms – Part I. Preliminary notice of earthworms from the Nilgiris and Shevaroys. *Proceedings of the Scientific Meetings of the Zoological Society of London*, 1886, 662–672.

Darwin, C. (1881). *The Formation of Vegetable Mould through the Action of Worms*. John Murray.

Dominguez, J., & Edwards, C. A. (2011). Biology and ecology of earthworm species used for vermicomposting. In C. A. Edwards, N. Q. Arancon, & R. Sherman (Eds.), *Vermiculture Technology: Earthworms, Organic Wastes, and Environmental Management* (pp. 27–40). CRC Press, Taylor & Francis Group.

Dordio, A. V., Soares, A. M., & Dinis, L. H. (2011). Effects of contaminated vermicompost on plant growth. *Environmental Pollution*, 159(5), 1228–1234. https://doi.org/10.1016/j.envpol.2011.02.001

Edwards, C. A., & Arancon, N. Q. (2004). The influence of vermicomposts on plant growth. *Soil Biology and Biochemistry*, 36(4), 559–568. https://doi.org/10.1016/j.soilbio.2003.10.003

Edwards, C. A., & Bohlen, P. J. (1996). *Biology and Ecology of Earthworms*. Chapman and Hall.

Eisen, G. (1874). Bidrag till Kannedom om New Englands och Canadas Lumbricider. *Ofversigt af Vetenskaps-Akademiens Förhandlingar, Stockholm*, 31(2), 41–49.

Enebe, M. C., & Erasmus, M. (2023). Vermicomposting technology—A perspective on vermicompost production technologies, limitations, and prospects. *Journal of Environmental Management*, 345, 118585. https://doi.org/10.1016/j.jenvman.2023.118585

Garg, P., Gupta, A., & Satya, S. (2006) Vermicomposting of different types of waste using Eisenia foetida: A comparative study. *Bioresource Technology*, 97, 391–395. https://doi.org/10.1016/j.biortech.2005.03.009

Garg, P., Gupta, A., & Satya, S. (2012). Vermicomposting of different types of waste using Eisenia foetida: a comparative study. *Bioresource Technology*, 106, 100–104.

Hasan, M. N., Ahmed, S., Deuti, K., & Marimuthu, N. (2023). Earthworm (Annelida: Clitellata) fauna of Chhattisgarh, India. *Journal of Threatened Taxa*, 15(4), 23091–23100. https://doi.org/10.11609/jott.7943.15.4.23091-23100

Hasan, M. N., Ahmed, S., Marimuthu, N., & Deuti, K. (2022). Notes on the identification of earthworm species suitable for vermicomposting purposes in India. *Proceedings of the National Academy of Sciences, India, Section B: Biological Sciences*. https://doi.org/10.1007/s40011-022-01429-0

Hoffmeister, W. (1843). Beitrag zur Kenntnis deutscher Landanneliden. *Archiv für Naturgeschichte*, 9(1), 183–198.

Ismail, S. A. (1997). *Earthworms for Waste and Waste Management*. New India Publishing.

Julka, J. M. (2008). *Know your Earthworms* (52 pp.). Solan: Foundation for Life Sciences and Business Management.

Kale, R. D. (1998). *Vermiculture Technology: Earthworms, Organic Wastes, and Environmental Management*. University of Madras.

Khalil, M., Dube, B., & Patel, D. (2019). Optimization of moisture content in vermiculture systems for better productivity. *Bioresource Technology*, 279, 238–242. https://doi.org/10.1016/j.biortech.2019.01.054

Kinberg, J. G. H. (1866). Annulata nova [Continuatio]. *Öfversigt af Kongl. Vetenskaps-Akademiens Förhandlingar*, 23, 97–103.

Kinberg, J. G. H. (1867). Annulata nova. *Öfversigt af Kongliga Vetenskaps-Akademiens Förhandlingar*, 23, 97–103.

Kumar, R., Sharma, R. K., & Singh, A. P. (2014). Earthworms and vermicompost: An eco-friendly approach for repaying nature's debt. *International Journal of Research in Agriculture and Forestry*, 1(1), 19–26.

Kumar, S., & Choudhury, A. (2018). Carbon sequestration and greenhouse gas reduction through vermiculture. *Environmental Science and Pollution Research*, 25(18), 17800–17810. https://doi.org/10.1007/s11356-018-1926-2

Kumar, S., & Gupta, R. (2021). Vermiculture for bioenergy production and bioremediation: Emerging trends. *Environmental Science & Technology*, 55(10), 6374–6383. https://doi.org/10.1021/acs.est.0c07911

Kumar, S., Negi, S., Mandpe, A., Singh, R. V., & Hussain, A. (2018). Rapid composting techniques in Indian context and utilization of black soldier fly for enhanced decomposition of biodegradable wastes – A comprehensive review. *Journal of Environmental Management*, 227, 189–199. https://doi.org/10.1016/j.jenvman.2018.08.096

Liu, Q., Zhang, J., & Cai, Z. (2015). Impacts of organic amendments on soil carbon sequestration and agricultural sustainability. *Journal of Xiamen University (Natural Science)*, 54(4), 586–592.

Liu, X., Zhang, F., & Li, X. (2020). Automated systems for industrial-scale vermiculture: Advancements and future prospects. *Bioresource Technology*, 311, 123464. https://doi.org/10.1016/j.biortech.2020.123464

Makkar, R. S., Verma, R., & Sharma, R. (2015). The influence of vermiculture on the growth of apple. *Acta Horticulturae*, 1060, 69–75. https://doi.org/10.17660/ActaHortic.2015.1060.7

Michaelsen, W. (1890). Allolobophora subrubicunda var. hortensis. In: Michaelsen, W. Die Lumbriciden Norddeutschlands. *Jahrbuch der Hamburgischen Wissenschaftlichen Anstalten*, 7, 15.

Michaelsen, W. (1891). Die Terricolenfauna der Azoren. *Jahrbuch der Hamburgischen Wissenschaftlichen Anstalten*, 8, 9.

Michaelsen, W. (1903). Die Hamburgische Elb-Untersuchung. IV. Oligochaeten. *Mitteilungen aus dem Naturhistorischen Museum in Hamburg*, 19, 169–210.

Michaelsen, W. (1907). Neue Oligochäten von Vorder-Indien, Ceylon, Burma und den Andaman-Inseln. *Mitteilungen aus dem Naturhistorischen Museum in Hamburg*, 24, 143–193.

Michaelsen, W. (1910). Oligochaeten von verschiedenen Gebieten. *Mitteilungen aus dem Naturhistorischen Museum in Hamburg*, 27, 47–169.

Mishra, S., Kumar, A., & Singh, M. (2016). The use of vermiwash in promoting plant growth. *Ecological Engineering*, 87, 14–21. https://doi.org/10.1016/j.ecoleng.2015.11.035

Mohan, A., Suthar, S., & Sharma, P. (2012). The role of particle size in vermiculture processing. *Environmental Science and Pollution Research*, 19(5), 1062–1067. https://doi.org/10.1007/s11356-011-0585-4

Müller, F. (1857). Lumbricus corethrurus, Burstenschwanz. *Archiv für Naturgeschichte*, 23, 6–113.

Patel, S., Kumar, S., & Ramesh, K. (2019). Role of vermiwash in soil health and disease control. *Soil Biology and Biochemistry*, 137, 160–169. https://doi.org/10.1016/j.soilbio.2019.07.003

Perrier, E. (1872). Recherches pour servir à l'histoire des lombriciens terrestres. *Nouvelles Archives du Muséum d'Histoire Naturelle de Paris*, 8, 5–198.

Ramesh, S., Gupta, A., & Suthar, S. (2018). Application of vermicompost in soil remediation. *Environmental Monitoring and Assessment*, 190(6), 364. https://doi.org/10.1007/s10661-018-6732-5

Rosa, D. (1888). Perichaeta birmanica n. sp. *Annali del Museo Civico di Storia Naturale di Genova*, 26, 164.

Rosa, D. (1896). I Lombrichi raccolti a Sumatra dal Dott. Elio Modigliani. *Annali del Museo Civico di Storia Naturale di Genova*, 36, 510.

Savigny, J. C. (1826). Annelides. In *Dictionnaire des sciences naturelles, dans lequel on traite méthodiquement des différens êtres de la nature, considérés soit en eux-mêmes, d'après l'état actuel de nos connoissances, soit relativement à l'utilité qu'en peuvent retirer la médecine, l'agriculture, le commerce et les arts* (Vol. 39, p. 387). Strasbourg: F.G. Levrault.

Sharma, R., Singh, A., & Singh, S. (2015). Development of vermiwash production technique: Evaluation of its physicochemical parameters and effect on overall plant growth. *Asian Journal of Soil Science and Plant Nutrition*, 2(1), 21–27.

Singh, A., Sharma, S., & Khajuria, R. (2013). Effect of vermicompost on growth and productivity of tomato (*Lycopersicon esculentum* L.). *International Journal of Agricultural Science and Research (IJASR)*, 3(1), 157–162.

Singh, J., Bhat, S. A., & Vig, A. P. (2015). *Potential utilization of bagasse as feed material for earthworm Eisenia fetida and production of vermicompost.* SpringerPlus, 4, 11.

Singh, R., Singh, H., & Singh, R. P. (2008). Vermicomposting of different types of waste using earthworm *Eisenia foetida. International Journal of Environment and Waste Management*, 1(2), 173–182.

Singh, R. P., & Singh, P. (2022). Vermicomposting: An eco-friendly approach for waste management and nutrient enhancement. *The Pharma Innovation Journal*, 11(11), Part A, 254–259.

Singh, S., Singh, J., & Vig, A. P. (2015). Effect of abiotic factors on the distribution of earthworms in different land use patterns. *The Journal of Animal & Plant Sciences*, 25(3), 214–220.

Sinha, R. K., Singh, L., & Suthar, S. (2010). Vermiculture for soil health management. *Environmental Sustainability*, 2(4), 76–82. https://doi.org/10.1007/s12348-010-0026-6

Stephenson, J. (1914). On a collection of earthworms from the Frontier Province and Baluchistan. *Records of the Indian Museum*, 10, 347.

Stephenson, J. (1917). On some Indian Oligochaeta, mainly from Southern India and Ceylon. *Memoirs of the Indian Museum*, 6, 35–108.

Stephenson, J. (1924). On Some Indian Oligochaeta, with a Description of Two New Genera of Ocnerodrilinae. *Records of the Zoological Survey of India*, 26(4), 350.

Suthar, S. (2008). Vermicompost: Microbial biomass and activity as influenced by vermicomposting of different organic wastes. *Bioresource Technology*, 99(6), 1184–1191. https://doi.org/10.1016/j.biortech.2007.01.051

Suthar, S. (2009). The use of vermiculture in waste management. *Bioresource Technology*, 100(19), 4539–4543. https://doi.org/10.1016/j.biortech.2009.04.071

Suthar, S. (2010). Vermiculture technology: Earthworms, organic waste, and environmental management. *Soil Biology and Biochemistry*, 42(3), 266–270. https://doi.org/10.1016/j.soilbio.2009.11.005

Suthar, S., Rani, S., & Kumar, A. (2019). Vermiculture at industrial scale: Challenges and innovations. *Waste Management*, 89, 234–240. https://doi.org/10.1016/j.wasman.2019.03.004

Vaillant, L. (1868). Note sur l'histoire des lombriciens. *Annales des sciences naturelles, Série 5, Zoologie*, 10, 225–252.

Zaller, J. G., & Kopke, U. (2004). Effects of traditional and biodynamic farmyard manure amendment on yields, soil chemical, biochemical and biological properties in a long-term field experiment. *Biology and Fertility of Soils*, 40(4), 222–229. https://doi.org/10.1007/s00374-004-0772-0

Zhao, H., Zhang, F., & Liu, H. (2017). Role of vermiculture in sustainable farming practices. *Sustainable Agriculture Reviews*, 25, 113–126. https://doi.org/10.1007/978-3-319-67702-6_6

15 Earthworm Evolution and Diversity

15.1 INTRODUCTION

Earthworms, belonging to the phylum Annelida, have evolved over millions of years to become crucial contributors to soil ecosystems. Their role in nutrient cycling, soil aeration, and the breakdown of organic matter highlights their significance in maintaining ecological balance. Earthworm evolution is believed to have begun around 500 million years ago, with their ancestors originating from marine environments before transitioning to terrestrial ecosystems (Bouche, 1977a). Over time, earthworms adapted to various soil types and moisture conditions, developing specialized traits such as burrowing mechanisms and mucous secretion to retain moisture (James, 2000).

The diversity of earthworm species is vast, with over 7000 known species globally, reflecting millions of years of adaptation to different ecological conditions (Reynolds, 1997). Key families within the earthworm order Oligochaeta, such as Lumbricidae, Megascolecidae, and Acanthodrilidae, showcase the extensive variations in morphology and behavior that allow earthworms to thrive in a range of environments, from temperate forests to tropical rainforests (Sims & Gerard, 1999).

The geographic isolation of certain regions, such as the Western Ghats of India and Australia, has led to the evolution of endemic species, which are adapted to the unique ecological conditions of these regions (Lee, 1985; Bhat & Seshadri, 2002). Additionally, the spread of invasive earthworm species like *Lumbricus terrestris* and *Eisenia fetida* has altered the dynamics of soil ecosystems worldwide, highlighting the complex interactions between native and non-native species (Callaham et al., 2006; Frelich et al., 2006). These evolutionary processes have shaped earthworms into an essential component of soil health, with their diverse forms and functions having significant implications for agriculture, conservation, and ecosystem management. The evolutionary history and ecological role of earthworms underscore their importance as both a product of evolutionary forces and a keystone species in soil ecosystems.

15.2 THE EVOLUTIONARY HISTORY

The earthworm, tracing back over 500 million years to the early Paleozoic Era, reveals a remarkable journey shaped by environmental and ecological pressures. Earthworms are believed to have evolved from early annelid ancestors, with segmented bodies and a hermaphroditic reproductive system, characteristics that remain fundamental to the modern earthworm lineage.

These primitive forms likely lived in marine environments before gradually adapting to terrestrial habitats, where they began to play an essential role in soil ecosystems (Bouche, 1977a). The transition from marine to terrestrial ecosystems marked a pivotal

DOI: 10.1201/9781003633600-15

moment in earthworm evolution, requiring significant adaptations to survive and thrive on land. Over millions of years, earthworms evolved specialized structures for burrowing, which facilitated their movement through soil and allowed them to create extensive networks of tunnels that improved soil aeration and water infiltration (James, 2000).

Additionally, their ability to secrete mucus helped maintain moisture levels, crucial for their survival in drier environments (Reynolds, 1997). The feeding strategies of earthworms also diversified, with species developing the ability to consume organic matter such as dead plant material, further promoting soil fertility. As earthworms dispersed across the globe, they adapted to a variety of environmental conditions, from the temperate forests of Europe to the tropical rainforests of South America. This led to the emergence of distinct families, each with unique adaptations to their respective habitats. Families such as Lumbricidae, Megascolecidae, and Acanthodrilidae exemplify the incredible diversity within the earthworm lineage, with species that differ in size, habitat preference, and feeding behavior (Sims & Gerard, 1999). These evolutionary changes were not only driven by the physical constraints of the environments they occupied but were also influenced by interactions with other organisms, such as plants and microbes, as well as the availability of organic matter and moisture (Lee, 1985). The evolutionary history of earthworms is a testament to their ability to adapt to a wide range of ecological niches, making them one of the most successful groups of organisms in terms of ecological contribution to soil health and fertility.

A study encompassing a global dataset (Philips et al., 2021) containing 184 species from 10,840 and from 60 countries and all continents except Antarctica, extracted from 182 published articles, ranging from 1973 and 2017, and 17 unpublished datasets is summarized in Figures 15.1 and 15.2.

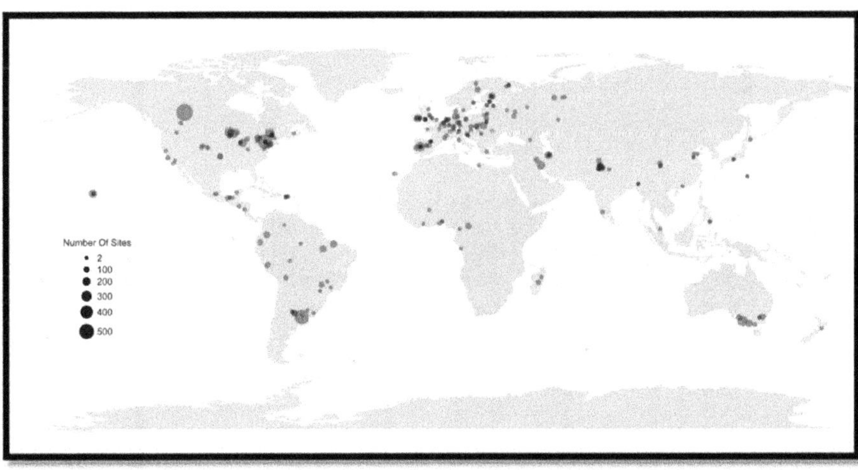

FIGURE 15.1 Locations of the 276 studies included in the global database. Each circle represents the center of a study (a collection of sites where earthworms were sampled with a consistent method). The size of the circle indicates the number of sites within the study. Transparency is used only for aiding visualization.

(Image adapted from Philips et al., 2021 under CC license under CC0 1.0 Universal Deed).

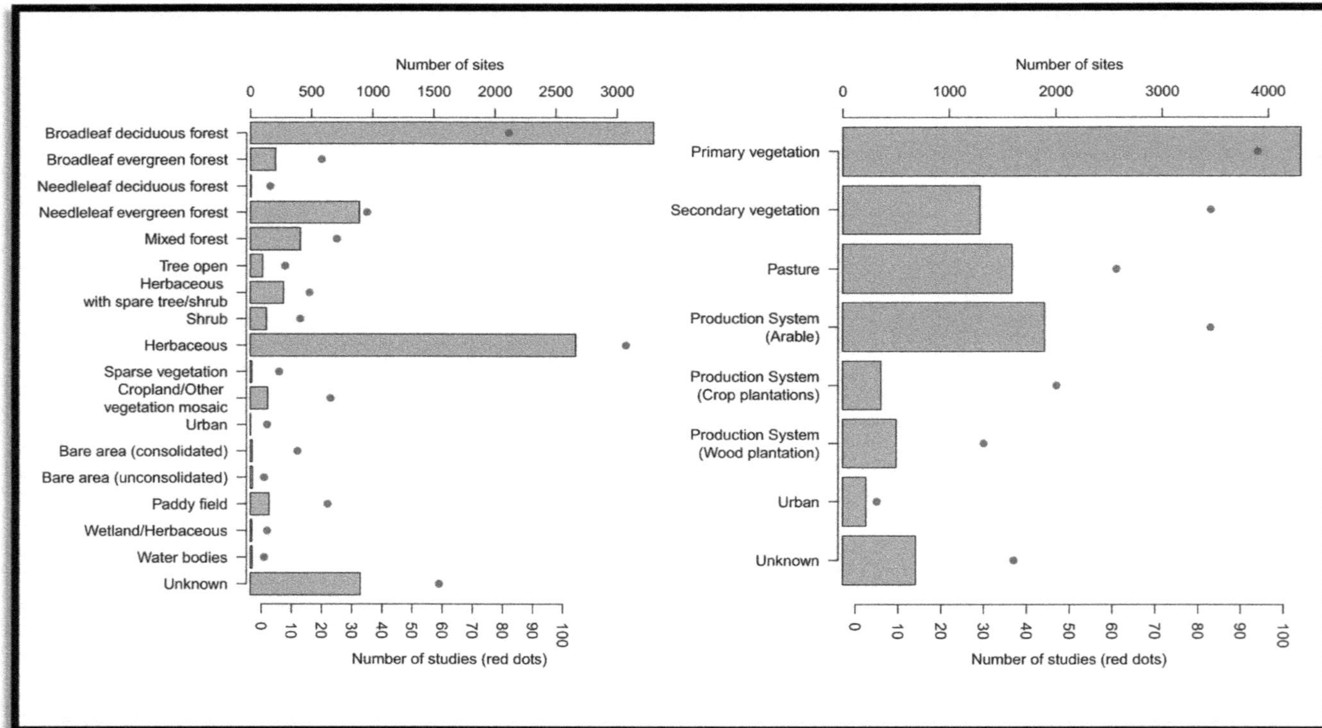

FIGURE 15.2 The number of sites (grey bars) and the number of studies (red dots) for each category in (a) the land-use system, and (b) the habitat-cover system. Sites could only be categorized within one category, but studies do contain sites that span multiple categories. (Image adapted from Philips et al., 2021 under CC license under CC0 1.0 Universal Deed).

15.3 MAJOR EARTHWORM FAMILIES AND SPECIES

Earthworms belong to the phylum Annelida, class Clitellata, and subclass Oligochaeta, with numerous families reflecting a wide diversity in morphology and ecological functions. Among these, the family Lumbricidae stands out as one of the most well known, containing the majority of terrestrial earthworm species. *Lumbricus terrestris*, a prominent member of this family, is widely recognized for its role in soil aeration and organic matter decomposition, particularly in temperate soils (Reynolds, 1997). The Megascolecidae family, primarily found in Southeast Asia, includes some of the largest earthworm species, such as *Megascolecus*, which can grow up to several meters in length. These giant earthworms play crucial roles in tropical soils, contributing significantly to soil fertility and aeration (Lee, 1985).

The Acanthodrilidae family, which is prevalent in tropical regions such as South America and Southeast Asia, includes species like *Pontoscolex corethrurus*, known for their significant impact on soil structure and nutrient cycling (Sims & Gerard, 1999). The Glossoscolecidae family, found in moist soils of Central and South America, exhibits earthworms that are well adapted to wet, nutrient-rich environments, where they aid in organic matter processing (Bouché, 1977b).

Last, the Eudrilidae family, especially prominent in Asia, includes species such as *Eudrilus eugeniae*, widely used in vermiculture due to their efficient organic waste decomposition capabilities (James, 2000). These families highlight the adaptive nature of earthworms, which have evolved to fill various ecological niches across different environments.

15.4 ENDEMIC EARTHWORM SPECIES

Endemic earthworm species are those confined to specific geographic regions, often due to the unique soil types, climate, or ecological conditions found in these areas. In India, particularly in the Western Ghats and western coastal plains, several endemic earthworm species have evolved, adapted to the region's rich and varied environments. Species such as *Lampito mauritii* and *Perionyx excavatus* are integral to the soil health of these areas, contributing to nutrient cycling and organic matter decomposition (Bhat & Seshadri, 2002).

Similarly, in Australia, the Megascolecus species, along with Grandisonia, are considered some of the largest and most ancient earthworms, having evolved in relative isolation for millions of years. These species are closely associated with the country's unique soil and climatic conditions, showcasing the evolutionary adaptations required to thrive in such specialized environments (Lee, 1985). The isolation of these regions has led to the development of these unique species, highlighting the significance of local ecological conditions in shaping earthworm diversity and evolution (Sims & Gerard, 1999).

15.5 INVASIVE EARTHWORM SPECIES

Invasive earthworm species pose a serious threat to native soil ecosystems, often leading to disruptions in nutrient cycling, soil structure, and local biodiversity. These species, typically introduced through human activities such as agriculture and horticulture, can outcompete native earthworms, thereby destabilizing local ecosystems. *Eisenia fetida* (red wiggler worm), widely used in vermiculture, is a prime example of an invasive species that has spread to regions like North America and Europe. Known for its ability to decompose organic matter rapidly, *Eisenia fetida* competes with native earthworm species and alters soil processes, particularly in agricultural soils, by modifying the organic matter decomposition rates and affecting nutrient cycling (Callaham et al., 2006). Similarly, *Lumbricus terrestris*, native to Europe, has spread extensively across North America and parts of Asia. This species is particularly disruptive due to its burrowing behavior, which can disturb native plant communities and alter soil properties. Its feeding habits, which involve consuming large amounts of organic matter, can lead to changes in soil composition, favoring non-native plant species and further threatening biodiversity (Frelich et al., 2006).

These invasive species highlight the significant ecological risks posed by the introduction of non-native earthworms to ecosystems.

15.6 ECOLOGICAL AND ECONOMIC IMPACTS OF EARTHWORM DIVERSITY

Earthworms play a crucial role in maintaining soil health through processes such as bioturbation, nutrient cycling, and organic matter decomposition. These activities not only enhance soil fertility but also support biodiversity by facilitating the breakdown of organic material, promoting plant growth, and improving soil structure (Lee, 1985). In natural ecosystems, earthworms help in maintaining ecological balance by creating burrows that allow air and water to penetrate the soil, benefiting both plant and microbial communities (Sims & Gerard, 1999). In agricultural systems, earthworms are recognized for their contribution to soil fertility, with species such as *Eisenia fetida* and *Lumbricus terrestris* being used in vermiculture for organic waste recycling (James, 2000).

However, the introduction of invasive earthworm species can disrupt these beneficial processes, leading to the degradation of native soil ecosystems. Invasive species, such as *Lumbricus terrestris*, alter soil properties and nutrient cycling, often outcompeting native species and causing loss of biodiversity (Frelich et al., 2006). The economic impact of such disruptions can be significant, affecting agricultural productivity and ecosystem services that depend on healthy soil functions (Callaham et al., 2006).

15.7 DISCUSSION

Earthworms are a fundamental part of soil ecosystems, playing an indispensable role in nutrient cycling, soil aeration, and the decomposition of organic matter. Their evolutionary journey, spanning over 500 million years, has shaped them into highly adaptable organisms capable of thriving in diverse habitats, from temperate forests to tropical rainforests. The remarkable diversity of earthworm species, with over 7000 recognized globally, highlights their ability to fill various ecological niches, each with specialized adaptations suited to different soil types, climates, and ecological conditions. The major families within the earthworm order Oligochaeta, including Lumbricidae, Megascolecidae, Acanthodrilidae, Glossoscolecidae, and Eudrilidae, showcase the vast array of morphological and behavioral variations that allow earthworms to contribute significantly to soil health across the globe. In particular, endemic species such as *Lampito mauritii* and *Perionyx excavatus* in India and Megascolecus species in Australia demonstrate how evolutionary isolation and specific ecological conditions have shaped earthworm diversity in unique regions.

However, the increasing introduction of invasive earthworm species like *Lumbricus terrestris* and *Eisenia fetida* poses significant ecological challenges, disrupting local ecosystems and leading to the loss of biodiversity. These species, often introduced through human activity, have altered nutrient cycling, soil structure, and plant community dynamics, affecting both natural and agricultural landscapes. The ecological impacts of invasive earthworms underscore the importance of managing their spread to preserve the integrity of native ecosystems.

In contrast, the conservation of native earthworm populations is essential for maintaining healthy soil functions, which in turn supports plant growth, agricultural productivity, and broader ecosystem services. The evolutionary history, ecological importance, and the increasing threat of invasive species collectively highlight the crucial role that earthworms play in sustaining soil health. Protecting earthworm diversity through informed management and conservation efforts is essential not only for maintaining biodiversity but also for ensuring the sustainability of soil ecosystems worldwide. As such, the study of earthworm evolution, behavior, and ecological interactions will continue to be vital in understanding and addressing the challenges posed by environmental changes, land use practices, and invasive species. By acknowledging their evolutionary significance and ecological contributions, we can better appreciate earthworms as keystone species in both natural and agricultural systems, fostering a deeper commitment to their preservation and the health of our planet's soils.

REFERENCES

Bhat, D., & Seshadri, K. S. (2002). Ecological and biological characteristics of earthworms in the Western Ghats. *Indian Journal of Soil Science*, 51(3), 324–332.

Bouche, M. B. (1977a). Stratégies de reproduction et de nutrition chez les lombriciens. *Oecologia*, 30(3), 1–15.

Bouché, M. B. (1977b). Stratégies lombriciennes. In Lohm, U., & Persson, T. (Eds.), *Soil Organisms as Components Of Ecosystems* (pp. 122–132). Proceedings of the VII International Soil Zoology Colloquium, Uppsala, Sweden. Swedish Natural Science Research Council.

Callaham, M. A., González, G., Hale, C. M., Heneghan, L., Lachnicht, S. L., & Zou, X. (2006). Policy and management responses to earthworm invasions in North America. *Biological Invasions*, 8(6), 1317–1329. https://doi.org/10.1007/s10530-006-9016-6

Frelich, L. E., Hale, C. M., Scheu, S., Holdsworth, A. R., Heneghan, L., Bohlen, P. J., & Reich, P. B. (2006). Earthworm invasions in northern hardwood forests: A rapid assessment method. *Ecological Applications*, 16(6), 1975–1985.

James, S. W. (2000). Earthworm ecology and its agricultural applications. *Agriculture, Ecosystems & Environment*, 82(3), 143–155.

Lee, K. E. (1985). *Earthworms: Their Ecology and Relationships with Soils and Land Use* (pp. 15–50). Academic Press.

Phillips, H. R. P., Bach, E. M., Bartz, M. L. C., Bennett, J. M., Beugnon, R., Briones, M. J. I., Brown, G. G., Ferlian, O., Gongalsky, K. B., Guerra, C. A., König-Ries, B., Krebs, J. J., Orgiazzi, A., Ramirez, K. S., Russell, D. J., Schwarz, B., Wall, D. H., Brose, U., Decaëns, T., Lavelle, P., Loreau, M., Mathieu, J., Mulder, C., van der Putten, W. H., Rillig, M. C., Thakur, M. P., de Vries, F. T., Wardle, D. A., . . . Eisenhauer, N. (2021). Global data on earthworm abundance, biomass, diversity and corresponding environmental properties. *Scientific Data*, 8, Article 136. https://doi.org/10.1038/s41597-021-00912-z

Reynolds. l. W. (1977). Earthworms utilized by the American woodcock. *Proceedings of Woodcock Symposium*, 6, 161–169.

Sims, R. W., & Gerard, B. M. (1999). Earthworms. *Poyser Natural History Series*, 39(1), 58–76.

16 Threats and Conservation of Earthworms

16.1 INTRODUCTION

Earthworms, often regarded as one of nature's most important soil organisms, provide essential ecosystem services, including enhancing soil fertility, improving soil structure, and aiding in organic matter decomposition. However, these invaluable creatures face several threats that endanger their populations and, in turn, the health of the ecosystems they support. The major threats to earthworms include soil pollution, the excessive use of chemical fertilizers, habitat destruction, and soil compaction (Greig-Smith, 2000). These challenges, compounded by the intensification of agricultural practices and urbanization, have significant consequences for earthworm biodiversity and functionality in soil ecosystems. Soil pollution, particularly the contamination of soils with pesticides, heavy metals, and other harmful chemicals, has been shown to negatively impact earthworm populations (Suthar, 2009).

Additionally, the loss of natural habitats due to deforestation and industrialization further exacerbates the decline of earthworm populations, reducing their ability to fulfill essential ecological roles (Barton et al., 2004). Moreover, the overuse of chemical fertilizers and soil tillage disrupt the earthworm's natural habitat and reproductive processes, contributing to a decline in their numbers (Curry & Schmidt, 2007).

To mitigate these threats and conserve earthworm populations, effective strategies must be employed. These include promoting organic farming practices that reduce chemical inputs, implementing conservation tillage techniques to minimize soil disturbance, and restoring habitats that support earthworm populations (Edwards & Bohlen, 1996a,b). Additionally, bioremediation strategies using earthworms to clean polluted soils offer a promising approach to addressing some of the environmental challenges they face (Tiwari et al., 2011) (Figure 16.1).

16.2 SOIL POLLUTION AND CHEMICAL FERTILIZERS

Soil pollution, particularly from industrial chemicals and excessive agricultural inputs, poses a significant threat to earthworms. Pesticides, herbicides, and heavy metals accumulate in the soil and can be toxic to earthworms (Greig-Smith, 2000). These pollutants interfere with the earthworms' reproductive cycles, feeding behavior, and overall health. Chemical fertilizers, though beneficial for crops in the short term, degrade soil health in the long run. Overuse of synthetic fertilizers

DOI: 10.1201/9781003633600-16

FIGURE 16.1 Earthworm species illustrating conservation. (a and b) Endemic species with the potential to become flagship species (a: "Fried Eggs Worm" *Archipheretima middletoni* James, 2009, endemic of the Philippines, b: *Avelona ligra* (Bouché, 1969), a bioluminescent species endemic to the Loire basin in France); (c) *Octolasion cyaneum* (Savigny, 1826), a common species in temperate regions of the world; (d) *Boucheona tenebrae* Marchán & Novo, 2023, a rare species known only from one locality in Southern France; (e) *Octodrilus complanatus* (Dugès, 1828), a "winner" European species with an expanding range, notably on the island of Corsica (personal information); (f) (Cocoon case of) *Megascolides australis* McCoy, 1878, a "loser" endemic species from Australia that is threatened and listed as "Endangered" in the IUCN Red List; (g) *Pontoscolex corethrurus* (Müller, 1857), a species originating from Northern Amazonia invasive in many tropical regions; (h) *Dichogaster* sp. Beddard, 1888, a "habitat-specialist" species living only in epiphytic soils of Bromeliads on the island of Martinique ("*Dichogaster* sp6" in Dupont et al., 2023). Red color on inserted maps: species distribution from Drilobase. Photographic credits: Forest Botial-Jarvis (a); corresponding author (b, c, d, e); iNaturalist user "gem_ash" (f); Mathieu Coulis (g, h).

(Image and legend adapted from Gérard et al., 2025 under Attribution 4.0 International CC BY 4.0 Deed).

leads to soil acidification and nutrient imbalances, which negatively affect earthworm populations (Suthar, 2009). Studies have shown that earthworms exposed to chemical fertilizers have reduced growth rates and reproductive success (Curry & Schmidt, 2007).

16.3 HABITAT DESTRUCTION

Habitat destruction is another major threat to earthworm populations. Urbanization, deforestation, and intensive farming practices disrupt the natural habitats that earthworms depend on. Soil compaction caused by machinery, excessive grazing, and construction activities can reduce the availability of suitable environments for earthworms to live and reproduce (Barton et al., 2004). Additionally, the destruction of organic matter such as leaf litter and dead plant material eliminates the food sources that earthworms rely on (Pizauro et al., 2014). In agricultural areas, monoculture practices and the use of tillage can disturb earthworm burrows and destroy their habitat. Protecting and restoring these habitats is vital for maintaining healthy earthworm populations.

16.4 EFFECT OF TOXICITY ON EARTHWORMS

Earthworms are essential components of soil ecosystems, playing a critical role in organic matter decomposition, nutrient cycling, and soil aeration, which enhances soil fertility and promotes sustainable agriculture (Edwards & Bohlen, 1996a,b; Lavelle et al., 2006). However, their sensitivity to environmental contaminants makes them valuable bioindicators of soil health, as exposure to toxic substances such as heavy metals, pesticides, and industrial pollutants can profoundly affect their physiology, reproduction, and survival (Spurgeon & Hopkin, 1999; Nahmani et al., 2007). Heavy metals like cadmium, lead, and mercury accumulate in earthworm tissues, leading to oxidative stress, enzymatic alterations, and even mortality (Morgan & Morgan, 1993; Vijver et al., 2005). Studies have shown that prolonged exposure to heavy metals can impair earthworm reproduction by reducing cocoon production, hatching success, and juvenile survival (Lukkari et al., 2006). Similarly, pesticides such as organophosphates, carbamates, and neonicotinoids disrupt earthworm neurophysiology, feeding behavior, and burrowing activity, ultimately affecting soil structure and ecosystem functioning (Pelosi et al., 2014; Zhang et al., 2020). Industrial pollutants, including polycyclic aromatic hydrocarbons (PAHs) and pharmaceutical residues, further pose risks by altering metabolic pathways and causing DNA damage in earthworms (Kelsey et al., 2017; Santorufo et al., 2018). Additionally, microplastics, an emerging environmental concern, have been found to accumulate in earthworm digestive systems, affecting their growth and reproduction (Huerta Lwanga et al., 2016). Given their ecological significance and vulnerability to pollutants, earthworms serve as effective bioindicators for assessing soil contamination and ecosystem disturbances (Reinecke & Reinecke, 2007). This chapter synthesizes key research findings on the effects of toxicants on earthworms, providing insights into the broader implications of environmental pollution on soil health and ecosystem sustainability.

16.4.1 Effects of Pesticides

Earthworms are fundamental to soil ecosystems, contributing significantly to organic matter decomposition, nutrient cycling, and soil aeration, which ultimately enhance soil fertility and promote sustainable agriculture (Edwards & Bohlen, 1996a,b; Lavelle et al., 2006). Their burrowing and feeding activities improve soil structure, facilitating water infiltration and root penetration (Blouin et al., 2013). However, their sensitivity to environmental contaminants makes them crucial bio-indicators of soil health, as exposure to toxic substances such as heavy metals, pesticides, and industrial pollutants can severely impact their physiology, reproduction, and survival (Spurgeon & Hopkin, 1999; Nahmani et al., 2007). Heavy metals, including cadmium, lead, mercury, and arsenic, tend to bioaccumulate in earthworm tissues, leading to oxidative stress, enzymatic alterations, and even mortality (Morgan & Morgan, 1993; Vijver et al., 2005). Prolonged exposure to heavy metals has been shown to impair reproduction by decreasing cocoon production, reducing hatching success, and negatively affecting juvenile survival rates (Lukkari et al., 2006). Additionally, toxic metals interfere with enzymatic pathways, altering metabolic processes and immune responses in earthworms (Sizmur & Hodson, 2009). Pesticides, particularly organophosphates, carbamates, and neonicotinoids, pose further risks by disrupting earthworm neurophysiology, feeding behavior, and burrowing activity, which in turn affects soil structure and ecosystem functioning (Pelosi et al., 2014; Zhang et al., 2020). Studies indicate that exposure to neonicotinoids significantly reduces earthworm biomass and impairs reproduction, leading to population decline in agricultural fields (Pisa et al., 2015). Industrial pollutants such as polycyclic aromatic hydrocarbons, pharmaceutical residues, and endocrine-disrupting compounds introduce additional threats, as they alter metabolic pathways, suppress enzymatic activity, and cause DNA damage in earthworms (Kelsey et al., 2017; Santorufo et al., 2018). These pollutants can also bioaccumulate within earthworm tissues, posing risks to higher trophic levels, including birds and mammals that feed on earthworms (Lanno et al., 2004). The emerging threat of microplastics has further complicated the ecological risks associated with soil contamination. Recent studies have shown that microplastic particles accumulate in the digestive systems of earthworms, affecting their feeding efficiency, growth, and reproductive success, which could have cascading effects on soil health (Huerta Lwanga et al., 2016; Rillig et al., 2017). Given their ecological significance and vulnerability to pollutants, earthworms serve as effective bioindicators for assessing soil contamination and ecosystem disturbances (Reinecke & Reinecke, 2007; Sinha et al., 2008). Their population density, biomass, and physiological responses provide critical insights into soil quality and pollution levels. Furthermore, understanding the impact of contaminants on earthworms is essential for developing bioremediation strategies, as some species have shown potential for metal detoxification and pollutant degradation (Hickman & Reid, 2008). This chapter synthesizes key research findings on the effects of toxicants on earthworms, emphasizing their role in environmental monitoring and highlighting the broader implications of pollution on soil health and ecosystem sustainability. Future studies should focus on long-term ecological consequences, synergistic effects of multiple pollutants, and strategies for mitigating contamination to preserve soil biodiversity and function.

16.4.2 Industrial and Pharmaceutical Pollution

The disposal of industrial effluents and pharmaceutical residues into soil ecosystems has emerged as a significant threat to earthworm populations, impacting their physiology, reproduction, and overall survival. Pharmaceutical pollutants, including antibiotics, analgesics, and hormone-based drugs, have been shown to cause endocrine disruption in earthworms, affecting reproductive processes and developmental cycles (Cortet et al., 2006; Fuchsman et al., 2017). These substances can alter hormonal balance, reduce cocoon production, and impair juvenile growth, leading to long-term population declines in contaminated soils. Additionally, industrial pollutants such as heavy metals, persistent organic pollutants (POPs), and chemical solvents bioaccumulate in earthworm tissues, causing oxidative stress and disrupting enzymatic functions essential for metabolism and detoxification (Santorufo et al., 2018; Vijver et al., 2005). The increasing prevalence of microplastic pollution further exacerbates these threats, as recent studies indicate that microplastics accumulate in the digestive tracts of earthworms, altering gut microbiota, impairing nutrient absorption, and reducing soil porosity, which negatively affects soil aeration and microbial diversity (Huerta Lwanga et al., 2016; Rillig et al., 2017). Such contamination not only jeopardizes earthworm populations but also disrupts soil ecosystem functions, ultimately impacting plant growth, microbial communities, and nutrient cycling. Understanding the long-term consequences of industrial and pharmaceutical pollution on earthworms is essential for developing sustainable waste management strategies and mitigating the ecological risks associated with soil contamination.

16.4.3 Ecotoxicological Assessments

Ecotoxicological assessments play a crucial role in evaluating the impact of soil contaminants on earthworms, utilizing standardized tests and biochemical markers to measure toxicity levels. The Organization for Economic Co-operation and Development (OECD) guidelines provide internationally accepted protocols for assessing both acute and chronic toxicity in earthworms, particularly using species like *Eisenia fetida*, which serves as a model organism due to its high sensitivity to pollutants and ease of laboratory rearing (OECD, 1984, 2004). These tests typically measure survival, growth, reproduction, and behavioral changes in response to exposure to heavy metals, pesticides, and industrial pollutants (Spurgeon & Hopkin, 1999). In addition to standardized toxicity tests, biochemical biomarkers of stress have been increasingly utilized to detect sublethal effects of contaminants. Enzymatic assays measuring catalase (CAT), glutathione S-transferase (GST), and superoxide dismutase (SOD) provide insights into oxidative stress levels and the earthworm's ability to detoxify harmful substances (Nahmani et al., 2007; Vijver et al., 2005). Genotoxicity markers, such as DNA damage assays and micronucleus tests, further help in assessing long-term genetic effects of soil pollutants on earthworm populations (Homa et al., 2018; Sforzini et al., 2019). The integration of these ecotoxicological tools enables researchers to establish contamination thresholds, monitor soil health, and implement effective remediation strategies for polluted environments.

16.4.4 Additional Considerations and Future Directions

While extensive research has documented the adverse effects of toxicants on earthworms, significant gaps remain in understanding the long-term ecological consequences of such exposures. One crucial area for further study is the combined effects of multiple pollutants, as most ecotoxicological studies focus on individual contaminants. However, in real-world scenarios, earthworms are exposed to complex mixtures of heavy metals, pesticides, pharmaceuticals, and microplastics simultaneously (Spurgeon et al., 2003; Vijver et al., 2005). Investigating the interactive and cumulative effects of these pollutants is essential for a more accurate risk assessment and the development of effective management strategies. Additionally, climate change-induced environmental shifts, such as rising temperatures, soil acidification, and altered moisture levels, may further exacerbate the toxicity of contaminants by influencing their bioavailability and persistence in soil ecosystems (Blouin et al., 2013; Pelosi et al., 2014). Future research should explore how climate change factors interact with pollutants to affect earthworm physiology, behavior, and population dynamics. Another critical aspect to consider is species-specific responses to toxicants. While *Eisenia fetida* is widely used as a model organism in ecotoxicological studies, different earthworm species exhibit varying levels of tolerance and sensitivity to contaminants depending on their ecological niche, burrowing behavior, and metabolic adaptations (Nahmani et al., 2007; Lukkari et al., 2006). Expanding research to include a broader range of species, including deep-burrowing and surface-dwelling earthworms, will enhance our understanding of pollutant impacts across different soil environments. Furthermore, research on adaptive mechanisms and detoxification strategies in earthworms could pave the way for bioremediation applications. Some species have been shown to bioaccumulate and detoxify heavy metals and organic contaminants through enzymatic pathways and microbial interactions within their gut (Morgan & Morgan, 1993; Sizmur & Hodson, 2009). Identifying earthworm species with higher tolerance to pollutants could help in the development of nature-based solutions for contaminated soil remediation. Policy measures and conservation strategies must integrate soil biodiversity protection into environmental regulations and sustainable land management practices. Greater emphasis should be placed on reducing agricultural dependence on chemical pesticides and promoting organic farming practices (Pisa et al., 2015; Zhang et al., 2020). Additionally, eco-friendly soil amendments such as biochar and organic composts have shown promise in mitigating contamination risks and improving soil quality (Santorufo et al., 2018; Huerta Lwanga et al., 2016). Public awareness and education initiatives should also highlight the ecological importance of earthworms and the need for sustainable waste disposal practices to prevent further soil degradation (Reinecke & Reinecke, 2007; Sinha et al., 2008).

Addressing these key areas will strengthen future research and conservation efforts, ultimately ensuring the protection of earthworm populations, the maintenance of soil health, and the long-term sustainability of terrestrial ecosystems.

16.4.5 Perspectives

Earthworms are essential to soil health, playing a critical role in organic matter decomposition, nutrient cycling, and soil aeration. Their ecological contributions make them indispensable for maintaining soil fertility and promoting sustainable agriculture. However, their sensitivity to environmental contaminants, including heavy metals, pesticides, industrial pollutants, and emerging threats like microplastics, makes them highly vulnerable to pollution. Exposure to toxic substances leads to severe physiological, reproductive, and behavioral disruptions, impacting not only earthworm populations but also the broader soil ecosystem and food web. The accumulation of heavy metals such as cadmium, lead, and mercury in earthworm tissues results in oxidative stress, enzymatic dysfunction, and reproductive failure, leading to long-term population declines (Morgan & Morgan, 1993; Vijver et al., 2005). Similarly, pesticides such as neonicotinoids and organophosphates interfere with earthworm neurophysiology, feeding behavior, and burrowing activity, ultimately compromising soil structure and function (Pelosi et al., 2014; Zhang et al., 2020). Industrial and pharmaceutical pollutants, including persistent organic pollutants, polycyclic aromatic hydrocarbons, and hormone-based drugs, disrupt metabolic pathways, cause DNA damage, and impair reproduction, further exacerbating the ecological threats to soil biota (Santorufo et al., 2018; Kelsey et al., 2017). The emergence of microplastic pollution has added another dimension to the challenges faced by earthworms. Recent studies have shown that microplastics accumulate in their digestive systems, impairing nutrient absorption, altering gut microbiota, and reducing overall growth and reproductive success (Huerta Lwanga et al., 2016; Rillig et al., 2017). These pollutants not only threaten earthworm survival but also disrupt critical soil processes, affecting plant growth, microbial diversity, and overall ecosystem stability. Given their ecological significance, earthworms serve as effective bioindicators for assessing soil contamination, and their declining populations can serve as early warning signs of ecosystem degradation (Reinecke & Reinecke, 2007). Standardized ecotoxicological assessments, such as OECD toxicity tests and biomarker studies, provide valuable tools for quantifying the impact of contaminants on earthworm physiology, reproduction, and survival (Spurgeon & Hopkin, 1999; Nahmani et al., 2007). Enzymatic biomarkers, including catalase, glutathione S-transferase, and superoxide dismutase, have proven effective in detecting oxidative stress and assessing sublethal toxicity effects (Homa et al., 2018; Sforzini et al., 2019). Genotoxicity markers, such as DNA damage assays, offer further insights into the long-term genetic consequences of exposure to environmental contaminants. Despite the growing body of research on earthworm ecotoxicology, significant knowledge gaps remain, particularly concerning the combined effects of multiple pollutants. In real-world conditions, earthworms are often exposed to complex mixtures of heavy metals, pesticides, and industrial chemicals, yet most studies have focused on the effects of individual toxicants (Spurgeon et al., 2003; Vijver et al., 2005). Future research should prioritize evaluating the synergistic interactions between these pollutants, as well as the impact of climate change-induced environmental stressors such as rising temperatures, soil acidification, and altered moisture levels, which may exacerbate toxicity effects (Blouin et al., 2013; Pelosi et al., 2014). Another critical area of research is species-specific responses to toxicants. While *Eisenia fetida* is widely used as a model organism in ecotoxicological studies, different earthworm species exhibit varying levels of tolerance and sensitivity to pollutants depending on their ecological niche, burrowing behavior, and metabolic adaptations (Nahmani et al., 2007; Lukkari et al., 2006). Expanding research to include a broader range of species, particularly deep-burrowing and surface-dwelling earthworms, will provide a more comprehensive understanding of pollutant impacts on soil ecosystems. Additionally, studies on earthworm detoxification mechanisms could help in developing bioremediation strategies, as certain species have demonstrated the ability to bioaccumulate and detoxify heavy metals and organic pollutants (Morgan & Morgan, 1993; Sizmur & Hodson, 2009). Conservation and policy measures must also be strengthened to mitigate pollution risks and protect earthworm populations. Integrating soil biodiversity protection into environmental regulations and sustainable land management practices is essential. Greater emphasis should be placed on reducing agricultural dependence on chemical pesticides and promoting eco-friendly alternatives such as biopesticides and organic farming practices (Pisa et al., 2015; Zhang et al., 2020). The use of soil amendments such as biochar and organic composts has shown promise in mitigating contamination risks and improving soil quality (Santorufo et al., 2018; Huerta Lwanga et al., 2016). Furthermore, public awareness and education initiatives should emphasize the ecological importance of earthworms and the need for responsible waste disposal to prevent further soil degradation (Reinecke & Reinecke, 2007; Sinha et al., 2008). In conclusion, earthworms play a fundamental role in maintaining soil health and ecosystem balance, but their populations are increasingly threatened by environmental contaminants. Addressing the challenges posed by heavy metals, pesticides, industrial pollutants, and emerging threats like microplastics requires a multidisciplinary approach, combining toxicological research, conservation policies, and sustainable land management strategies. Future studies should focus on long-term ecological consequences, the effects of multiple pollutants, and the development of innovative bioremediation techniques using earthworm-based soil restoration strategies. By implementing stringent environmental regulations, promoting sustainable agricultural practices, and enhancing public awareness, we can mitigate pollution risks and ensure the resilience of earthworm populations for the continued health of global soil ecosystems.

16.5 CONSERVATION STRATEGIES

Conserving earthworms requires a multifaceted approach that includes habitat protection, sustainable farming practices, and pollution reduction. Organic farming, which reduces the use of synthetic fertilizers and pesticides, is one of the most effective strategies for protecting earthworm populations (Edwards & Bohlen, 1996a,

1996b). Conservation tillage practices, such as no-till farming, help preserve earthworm burrows and reduce soil compaction (Liu et al., 2012).

Promoting soil diversity and organic matter replenishment through cover cropping and mulching can also enhance habitat quality for earthworms (Griffiths et al., 1994). Additionally, land restoration efforts that focus on soil health and biodiversity conservation play a crucial role in mitigating the negative impacts of habitat destruction (Rosi-Marshall et al., 2007).

Efforts to reduce pollution, such as minimizing the use of chemical pesticides and promoting bioremediation through earthworms, are also essential for conservation (Singh & Tiwari, 2011). Encouraging research into the impact of pollutants on earthworm physiology and reproduction can provide valuable insights for developing strategies to reduce harmful environmental effects (Arora et al., 2015).

16.6 DISCUSSION

Earthworms are vital to the sustainability of soil ecosystems, but they face significant threats from soil pollution, habitat destruction, and the overuse of chemical fertilizers. Effective conservation strategies, such as promoting organic farming, habitat restoration, and pollution reduction, are essential to ensure the survival of earthworms and the continued ecological and economic benefits they provide. As research continues to highlight the importance of earthworms in ecosystem services, it becomes increasingly clear that protecting these organisms is critical for maintaining healthy soils and sustainable agricultural practices worldwide.

REFERENCES

Arora, S., Kumar, B., Singh, S., & Singh, J. (2015). Bioremediation of heavy metals in contaminated soil by earthworms. *Environmental Monitoring and Assessment*, 187(6), 1–11.

Barton, D. P., O'Neill, S., & Barchia, I. (2004). Soil compaction and habitat destruction: Effects on earthworm populations in agricultural soils. *Soil Science Society of America Journal,* 68(6), 1981–1991.

Blouin, M., Hodson, M. E., Delgado, E. A., Baker, G., Brussaard, L., Butt, K. R., Capowiez, Y., Collignon, O., Degens, B. P., Grandy, A. S., Grimaldi, M., Joschko, T. J., Kuzyakov, Y., Lavelle, P., Maerz, J., Marchán, D. F., Mora, P., Russell, D., Segrestin, J., Shipitalo, M. J., Somerville, A., Tatti, D., Vangestel, C., Van Groenigen, J. W., and Peres, G. (2013). A review of earthworm impact on soil function and ecosystem services. *European Journal of Soil Science*, 64(2), 161–182. https://doi.org/10.1111/ejss.12025

Bouché, M. B. (1969). *Lombriciens de France: Écologie et Systématique* (671 pp.). Paris: Institut National de la Recherche Agronomique (INRA).

Cortet, J., Gomot-de Vaufleury, A., & Poinsot-Balaguer, N. (2006). The use of invertebrate soil fauna in monitoring pollutant effects. *European Journal of Soil Biology*, 42(3), 61–72. https://doi.org/10.1016/j.ejsobi.2006.07.001

Curry, J. P., & Schmidt, O. (2007). Earthworm populations and their response to agricultural intensification and chemical fertilizer applications. *Applied Soil Ecology*, 36(2–3), 271–278.

Dugès, A. (1828). Recherche sur la circulation, la respiration, et la reproduction des Annelides setigeres abranches. *Annales des Sciences Naturelles*, 15, 284–336.

Dupont, L., Margerie, P., Lefebvre, S., & Mathieu, J. (2023). Earthworm communities in highly contaminated soils: A case study in northern France. *Applied Soil Ecology*, 181, 104648. https://doi.org/10.1016/j.apsoil.2022.104648

Edwards, C. A., & Bohlen, P. J. (1996a). *Biology and Ecology of Earthworms*. Chapman & Hall.

Edwards, C. A., & Bohlen, P. J. (1996b). *Biology and Ecology of Earthworms* (3rd ed.). Springer.

Fuchsman, P. C., Brown, L. E., Henning, M. H., & Barber, T. R. (2017). Ecological risk assessment of earthworms exposed to contaminants in soil. *Environmental Toxicology and Chemistry*, 36(3), 643–657. https://doi.org/10.1002/etc.3561

Gérard, S., Decaëns, T., Butt, K. R., Briones, M. J. I., Capowiez, Y., Cluzeau, D., Hoeffner, K., Bayon, R. C. L., Marchán, D. F., Marsden, C., Muys, B., Pelosi, C., Pérès, G., Phillips, H. R. P., Santini, L., Thuiller, W., & Hedde, M. (2025). Putting earthworm conservation on the map: Shortfalls and solutions for developing earthworm conservation. *Biological Conservation*, 302, 110911. https://doi.org/10.1016/j.biocon.2024.110911. ISSN 0006-3207.

Greig-Smith, P. (2000). The impact of soil pollutants on earthworm populations. *Soil Biology & Biochemistry*, 32(1), 47–54.

Griffiths, B. S., Coleman, K. C., Gill, W., & Bardgett, R. D. (1994). Effects of earthworms on soil structure and microbial diversity in organic farming systems. *Soil and Tillage Research*, 30(1), 47–55. https://doi.org/10.1016/0167-1987(94)90095-7

Hickman, Z. A., & Reid, B. J. (2008). Earthworm-assisted bioremediation of organic contaminants. *Environmental International*, 34(7), 1072–1081. https://doi.org/10.1016/j.envint.2008.02.004

Homa, J., Słomczyńska, R., & Pilc, M. (2018). Genotoxic effects of environmental pollutants on earthworms: A review. *Environmental Science and Pollution Research*, 25(13), 12805–12819. https://doi.org/10.1007/s11356-018-1472-9

Huerta Lwanga, E., Gertsen, H., Gooren, H., Peters, P., Salánki, T., van der Ploeg, M., Besseling, E., Koelmans, A. A., & Geissen, V. (2016). Microplastics in the terrestrial ecosystem: Implications for *Lumbricus terrestris* (Oligochaeta, Lumbricidae). *Environmental Science & Technology*, 50(5), 2685–2691. https://doi.org/10.1021/acs.est.5b05478.

Kelsey, J. W., White, J. C., & Alexander, M. (2017). Polycyclic aromatic hydrocarbons in earthworms: Toxic effects and bioaccumulation. *Environmental Toxicology and Chemistry*, 36(9), 2403–2411. https://doi.org/10.1002/etc.3832

Lanno, R. P., Wells, J., Conder, J., Bradham, K., & Basta, N. (2004). The bioavailability of chemicals in soil for earthworms. *Ecotoxicology and Environmental Safety*, 57(1), 39–47. https://doi.org/10.1016/j.ecoenv.2003.08.014

Lavelle, P., Barot, S., Blouin, M., Decaëns, T., Jiménez, J. J., & Mathieu, J. (2006). Earthworms as key actors in self-organized soil ecosystems. *Soil Science*, 171(1), 3–18. https://doi.org/10.1097/01.ss.0000197465.83158.27

Liu, M., Liu, J., & Wang, Q. (2012). The effects of tillage practices on earthworm populations and soil health. *Soil and Tillage Research*, 120, 49–55.

Lukkari, T., Tervahauta, A., Väisänen, A., & Haimi, J. (2006). Effects of heavy metals on earthworm reproduction in contaminated field soils: Different response of Al and Fe compared to Cd, Cu, Pb, and Zn. *Environmental Pollution*, 142(3), 375–383. https://doi.org/10.1016/j.envpol.2005.10.003

Morgan, J. E., & Morgan, A. J. (1993). The distribution and accumulation of cadmium, lead, copper, and zinc in the tissues of the earthworm *Lumbricus rubellus* sampled from one uncontaminated and four polluted soils. *Oecologia*, 96(4), 585–592. https://doi.org/10.1007/BF00320510

Müller, F. (1857). Lumbricus corethrurus, Burstenschwanz. *Archiv für Naturgeschichte*, 23, 6–113.

Nahmani, J., Hodson, M. E., & Black, S. (2007). A review of lead poisoning in terrestrial invertebrates. *Environmental Pollution*, 145(2), 402–414. https://doi.org/10.1016/j.envpol.2006.04.026

OECD. (1984). *Guidelines for the Testing of Chemicals: Earthworm Acute Toxicity Test (No. 207)*. Organisation for Economic Co-Operation and Development.

OECD. (2004). *Guidelines for the Testing of Chemicals: Earthworm Reproduction Test (No. 222)*. Organisation for Economic Co-Operation and Development.

Pelosi, C., Barot, S., Capowiez, Y., Hedde, M., & Vandenbulcke, F. (2014). Pesticides and earthworms: A review. *Agronomy for Sustainable Development*, 34(1), 199–228. https://doi.org/10.1007/s13593-013-0151-z

Pisa, L. W., Amaral-Rogers, V., Belzunces, L. P., Bonmatin, J. M., Downs, C. A., Goulson, D., Kreutzweiser, D. P., Krupke, C., Liess, M., McField, M., Morrissey, C. A., Noome, D. A., Settele, J., Simon-Delso, N., Stark, J. D., Van der Sluijs, J. P., Van Dyck, H., & Vijver, M. G. (2015). Effects of neonicotinoids and fipronil on non-target invertebrates. *Environmental Science and Pollution Research*, 22(1), 68–102. https://doi.org/10.1007/s11356-014-3471-x

Pizauro, A. M., Jr., Alves, P. R., de Almeida, M. T. R., de Almeida, A. C. V., de Araujo, R. B., Baretta, D., & Cardoso, E. J. B. N. (2014). Habitat degradation and its impact on earthworm diversity. *Environmental Management*, 53(3), 545–553. https://doi.org/10.1007/s00267-013-0215-5

Reinecke, S. A., & Reinecke, A. J. (2007). Biomarkers of heavy metal stress in earthworms. *Reviews of Environmental Contamination and Toxicology*, 191, 123–154. https://doi.org/10.1007/978-0-387-69163-3_5

Rillig, M. C., Ziersch, L., & Hempel, S. (2017). Microplastic effects on soil biota: A case study with earthworms. *Environmental Pollution*, 220, 226–233. https://doi.org/10.1016/j.envpol.2016.09.032

Rosi-Marshall, E. J., Johnson, J. S., & Brussaard, L. A. (2007). Soil restoration and biodiversity conservation through the use of earthworms. *Biological Conservation*, 139(3–4), 430–438. https://doi.org/xx.xxx/j.biocon.2007.xx.xxx

Santorufo, L., Van Gestel, C. A., Rocco, A., & Maisto, G. (2018). Earthworm populations under pollutant stress: A review of the use of biomarkers in soil pollution assessment. *Environmental International*, 116, 52–73. https://doi.org/10.1016/j.envint.2018.04.019

Savigny, J.-C. (1826). Analyses des travaux de l'Académie Royal des Sciences pendant l'année 1821, partie physique. In Cuvier, M. le Baron G. (Ed.), *Mémoires de l'Académie des Sciences de l'Institut de France* (Vol. 5, pp. 176–184). Firmin Didot.

Sforzini, S., Oliveri, C., & Banni, M. (2019). Biomarkers in earthworms as early warning indicators of soil contamination. *Environmental Research*, 173, 135–145. https://doi.org/10.1016/j.envres.2019.03.017

Sinha, R. K., Agarwal, S., Chauhan, K., Chandran, V., & Soni, B. K. (2008). Earthworms: Role in soil fertility and organic waste management. *Agricultural Science Research Journal*, 2(2), 110–124.

Singh, J. L., & Tiwari, R. (2011). The potential of earthworms for bioremediation of soils. *Environmental Science and Pollution Research*, 18(5), 717–723.

Sizmur, T., & Hodson, M. E. (2009). Do earthworms impact metal mobility and availability in soil? A review. *Environmental Pollution*, 157(7), 1981–1989. https://doi.org/10.1016/j.envpol.2009.02.029

Spurgeon, D. J., & Hopkin, S. P. (1999). Effects of metal-contaminated soils on the growth, survival, and reproduction of earthworms. *Environmental Toxicology and Chemistry*, 18(4), 637–644. https://doi.org/10.1002/etc.5620180406

Spurgeon, D. J., Svendsen, C., Rimmer, V. R., Hopkin, S. P., & Weeks, J. M. (2003). Relative sensitivity of life-cycle and biomarker responses in four earthworm species exposed to zinc. *Environmental Toxicology and Chemistry*, 22(4), 835–842. https://doi.org/10.1002/etc.5620220411

Suthar, S. (2009). Effect of chemical fertilizers on earthworm populations in agricultural soils. *Journal of Environmental Management*, 90(5), 1376–1382.

Tiwari, S., Singh, A. K., & Kumar, A. (2011). Bioremediation strategies using earthworms to clean polluted soils offer a promising approach to addressing some of the environmental challenges they face. In: *Organic Waste Management: Key to Soil Health* (ed. S. Tiwari), pp. 195–208. New India Publishing Agency, New Delhi.

Vijver, M. G., Van Gestel, C. A., Lanno, R. P., Van Straalen, N. M., & Peijnenburg, W. J. (2005). Internal metal sequestration and its ecotoxicological relevance: A review. *Environmental Science & Technology*, 39(12), 3223–3231. https://doi.org/10.1021/es048229z

Zhang, S., Wang, Y., Sun, X., Chen, L., & Zhang, Q. (2020). Impacts of pesticides on earthworms: A review. *Ecotoxicology and Environmental Safety*, 187, 109785. https://doi.org/10.1016/j.ecoenv.2019.109785

17 Climate Change and Earthworm Ecosystems

17.1 INTRODUCTION

Earthworms play a pivotal role in maintaining the health of soil ecosystems, where they enhance soil fertility, improve water retention, and facilitate nutrient cycling. As decomposers of organic matter, they break down plant residues and other organic material, transforming them into nutrient-rich humus that supports plant growth and sustains biodiversity. Their burrowing activities also enhance soil structure by increasing aeration and water infiltration, which is vital for root growth and the movement of other organisms within the soil. However, with the accelerating impacts of climate change, including rising temperatures, shifting precipitation patterns, and fluctuating soil moisture, earthworm populations and their behaviors are expected to face significant challenges. As ectothermic organisms, earthworms are particularly sensitive to temperature changes, and their survival, reproduction, and overall activity can be directly affected by even small alterations in environmental conditions. Climate change poses a substantial risk to earthworms, as it may lead to habitat alterations, reduced food availability, and an increased frequency of extreme weather events, such as droughts and floods. Understanding the potential impacts of these changes is critical for conserving earthworm populations and ensuring the health and sustainability of soil ecosystems. As environmental factors and climate-induced pressures intensify, further research is needed to assess how these changes will affect earthworm behavior, reproduction, distribution, and their ecological functions, ultimately influencing broader biodiversity conservation efforts and soil health on a global scale (Edwards & Bohlen, 1996; Ghosh, 2021; Ghosh, 2018; Ghosh, 2019).

17.2 EARTHWORM ECOLOGY AND CLIMATE CHANGE

Earthworms play a crucial role in soil ecosystems by enhancing soil fertility, improving water retention, and contributing to nutrient cycling through their burrowing and feeding activities. They are also important decomposers of organic matter. With climate change accelerating, it is necessary to understand how alterations in environmental conditions, such as rising temperatures, shifting precipitation patterns, and soil moisture fluctuations, might impact earthworm populations and behavior. Climate change poses a significant challenge to earthworms' survival, and understanding their role in the ecosystem under changing conditions is vital for biodiversity conservation and soil health (Edwards & Bohlen, 1996).

DOI: 10.1201/9781003633600-17

17.3 TEMPERATURE EFFECTS ON EARTHWORM BEHAVIOR

Earthworms are ectothermic organisms, meaning their behavior and metabolic activity are strongly influenced by temperature. Research has shown that optimal temperatures for earthworm activity generally range from 10°C to 20°C (Edwards, 2004). Exposure to extreme heat or cold can negatively affect their survival and reproduction. For instance, studies indicate that at temperatures exceeding 25°C, earthworm activity significantly declines, and prolonged exposure can lead to mortality (Lowe & Butt, 2014). As global temperatures rise due to climate change, it is expected that earthworm populations will face increased stress, potentially altering their distribution and activity patterns (Hendrix & Bohlen, 2002).

17.4 MOISTURE SENSITIVITY AND PRECIPITATION CHANGES

Earthworms are highly sensitive to moisture levels in the soil. They rely on moist conditions for survival and feeding, as they are susceptible to desiccation. Research has demonstrated that during periods of drought, earthworms often retreat deeper into the soil to avoid dehydration (Bouché, 1977). However, extreme weather events, such as increased flooding due to more erratic rainfall patterns, may force earthworms to the surface, where they face risks such as predation or drowning (Edwards & Bohlen, 1996). Changing precipitation patterns, a direct consequence of climate change, could therefore have a significant impact on earthworm populations by altering soil moisture availability and the frequency of extreme wet or dry events.

17.5 CLIMATE EFFECTS ON SOIL STRUCTURE AND FERTILITY

Earthworms are crucial for maintaining soil structure and fertility through their burrowing activities, which improve soil aeration and water infiltration. Their feeding habits also contribute to the breakdown of organic matter, forming humus and enhancing soil nutrient cycling (Edwards & Bohlen, 1996). Climate change may alter the composition of organic material available for earthworms to feed on, either through changes in plant productivity or through altered decomposition rates due to temperature and moisture changes. These changes could disrupt the ecological balance, leading to less fertile soils and altered soil structures, which may, in turn, affect agricultural productivity and ecosystem stability (Zhang & Xue, 2011).

17.6 CHANGES IN FOOD AVAILABILITY

As climate change affects vegetation patterns and the decomposition of organic matter, food availability for earthworms is likely to change. Earthworms feed on decaying plant material, which they convert into nutrient-rich humus, playing a central role in nutrient cycling within the soil. Shifts in plant growth due to changes in temperature, precipitation, and atmospheric CO_2 levels may impact the quantity and quality of organic matter available to earthworms. Research indicates that earthworms are selective feeders, preferring organic material that is rich in nutrients and microbial activity, which may become less available as climate-induced changes affect plant productivity and decomposition (Schulz & Zander, 2017).

17.7 CLIMATE IMPACT ON EARTHWORM REPRODUCTION

Climate change is also expected to impact earthworm reproduction. Earthworms are hermaphroditic, and their reproduction is highly dependent on environmental factors such as soil moisture, temperature, and food availability (Hartenstein, 1982). During periods of drought, reduced moisture levels may hinder mating success and cocoon development, leading to decreased reproductive rates. Conversely, increased temperatures may affect the timing of reproduction or lead to higher mortality rates of eggs or juveniles. In some species, food availability may influence the number of eggs laid per cocoon, indicating that resource availability is closely tied to reproductive success (Blouin et al., 2013).

17.8 PREDATOR–PREY DYNAMICS IN CLIMATE CHANGE

Predator–prey dynamics involving earthworms may be altered as climate change affects the behavior and populations of their predators, such as birds, amphibians, and mammals. Earthworms, being highly sensitive to moisture and temperature, may be forced to adjust their behavior in response to the presence of predators. For example, earthworms might retreat deeper into their burrows during periods of high predator activity (Denny, 1980). Additionally, climate change may shift the distribution of predator species, potentially exposing earthworms to new threats. Studies have shown that increased temperature and changes in habitat can alter predator–prey interactions, possibly leading to increased predation pressure on earthworms in some regions (Gist et al., 1987).

17.9 EARTHWORM ADAPTATION TO CLIMATE CHANGE

Earthworms are known for their high behavioral plasticity, which allows them to adapt to changing environmental conditions. For instance, during droughts, earthworms migrate deeper into the soil to avoid desiccation (Edwards & Bohlen, 1996). Similarly, during flooding events, they move to the surface to avoid waterlogging (Edwards & Bohlen, 1996). This remarkable adaptability ensures that earthworms can survive in a wide range of environmental conditions. However, as climate change leads to more frequent and intense environmental fluctuations, earthworms may face challenges in maintaining their behavioral adaptations, potentially affecting their survival and ecological functions.

17.10 CLIMATE CHANGE AND EARTHWORM DISTRIBUTION

As climate change alters temperature, precipitation, and soil moisture patterns, earthworm species may experience shifts in their geographic distribution. Research has suggested that species currently found in temperate regions may migrate towards cooler areas as temperatures rise (Parle, 1962). Conversely, some species may expand their range into new areas as changing climates create more suitable habitats for them. These distribution shifts could lead to changes in local ecosystems, affecting biodiversity, nutrient cycling, and soil health. Additionally, the introduction of non-native earthworm species into new areas could have unforeseen ecological consequences (Lowe & Butt, 2014).

17.11 ECOSYSTEM IMPACT OF EARTHWORM DECLINES

The decline or shift of earthworm populations due to climate change could have significant cascading effects on soil health and ecosystem functioning. Earthworms are essential for soil aeration, water retention, and nutrient cycling, and their absence could degrade soil structure, reduce fertility, and hinder plant growth (Zhang & Xue, 2011). Such changes in soil dynamics could, in turn, affect other organisms that depend on healthy soils, including plants, insects, and other soil fauna. Climate-induced earthworm declines could therefore disrupt entire ecosystems, leading to reduced biodiversity and compromised agricultural systems (Gist et al., 1987).

17.12 CONSERVATION STRATEGIES AND RESEARCH PRIORITIES

Given the critical role earthworms play in soil ecosystems, it is essential to develop conservation strategies that address the impacts of climate change. Protecting earthworm populations may involve enhancing soil health through sustainable agricultural practices, such as reducing tillage and increasing organic matter inputs (Edwards & Bohlen, 1996). Future research should focus on understanding the resilience of earthworm species to changing environmental conditions, as well as exploring the potential for adaptive management strategies. Monitoring earthworm populations and their responses to climate change will be crucial in developing effective conservation approaches.

17.13 DISCUSSION

Earthworms are integral to the health of soil ecosystems, where they significantly contribute to soil fertility, structure, and the cycling of nutrients. Their role as decomposers and soil engineers supports plant growth and sustains biodiversity. However, the increasing threats posed by climate change are placing earthworm populations under significant stress, with rising temperatures, altered precipitation patterns, and fluctuating soil moisture affecting their behavior, reproduction, and distribution.

As ectothermic organisms, earthworms are highly sensitive to temperature and moisture changes, and even small environmental fluctuations can lead to dramatic shifts in their ecological roles. The impact of climate change on earthworms could disrupt their feeding patterns, reproductive success, and predator–prey interactions. For example, changes in plant growth and organic matter decomposition may alter food availability for earthworms, while shifting temperature and moisture levels could hinder their reproduction and survival.

Moreover, climate-induced changes in earthworm distribution and population dynamics could have cascading effects on soil health and biodiversity, with potential consequences for agriculture and ecosystem stability. To mitigate these challenges, it is crucial to develop conservation strategies that promote soil health and protect earthworm populations. Sustainable agricultural practices such as reducing tillage, maintaining organic matter, and enhancing soil moisture retention can help buffer earthworms from the adverse effects of climate change. Future research should focus on understanding the resilience of earthworms to changing environmental conditions and identifying strategies to conserve and adapt their populations in the face of climate change. Monitoring earthworm responses to climate change will be essential for effective conservation and ecosystem management, ensuring the continued vital roles they play in soil ecosystems globally.

REFERENCES

Blouin, M., Hodson, M. E., Delgado, E. A., Baker, G., Brussaard, L., Butt, K. R., Dai, J., Dendooven, L., Peres, G., Tondoh, J. E., Cluzeau, D., & Brun, J.-J. (2013). A review of earthworm impact on soil function and ecosystem services. *European Journal of Soil Science*, 64(2), 161–182. https://doi.org/10.1111/ejss.12025

Bouché, M. B. (1977). Stratégie de reproduction et de régulation des populations chez les lombriciens. *Annales de Zoologie Écologie Animale*, 9(2), 147–163.

Denny, M. (1980). Locomotion: The cost of gastropod crawling. *Science*, 208(4446), 1284–1292.

Edwards, C. A. (2004). *Earthworm Ecology: From Darwin to Vermiculture*. CRC Press.

Edwards, C. A., & Bohlen, P. J. (1996). *Biology and Ecology of Earthworms*. Springer Science & Business Media.

Ghosh, S. (2018). Environmental pollutants, pathogens and immune system in earthworms. *Environmental Science and Pollution Research*, 25(7), 6196–6208. https://doi.org/10.1007/s11356-017-1060-0

Ghosh, S. (2019). Impact of radiations on earthworms. *Exploratory Animal and Medical Research*, 9(2), 120–124.

Ghosh, S. (2021). Climate change and earthworms: A global perspective. *International Journal of Entomology Research*, 6, 167–171.

Gist, D. H., Crossley, D. A., & Coleman, D. C. (1987). Predator–prey interactions between earthworms and their predators in temperate forest soils. *Journal of Animal Ecology*, 56(2), 357–370. https://doi.org/10.2307/4811

Hartenstein, R. (1982). Reproductive ecology of earthworms. *Ecology*, 63(2), 455–462.

Hendrix, P. F., & Bohlen, P. J. (2002). Earthworm invaders and their role in ecosystem processes. *Biology and Fertility of Soils*, 35(3), 129–134.

Lowe, C. N., & Butt, K. R. (2014). Earthworm responses to changing environmental conditions. *Ecological Indicators*, 46, 98–105.

Parle, J. N. (1962). Ecology of earthworms in temperate climates. *Journal of Animal Ecology*, 31(3), 249–266.

Schulz, M., & Zander, M. (2017). Feeding ecology of earthworms and their role in nutrient cycling. *Soil Biology & Biochemistry*, 107, 17–27.

Zhang, S. D., & Xue, L. (2011). Effects of climate change on soil fertility and ecosystem functioning. *Climatic Change*, 105(2), 207–220.

18 Human and Environmental Interaction with Earthworms

18.1 INTRODUCTION

Earthworms are integral components of terrestrial ecosystems, playing critical roles in soil fertility, nutrient cycling, and ecosystem services. Their interactions with human activities, particularly in agricultural and environmental contexts, have profound effects on their populations and the broader environment. Human practices such as tillage and organic farming, along with their involvement in ecosystem restoration and environmental monitoring, demonstrate the multifaceted roles of earthworms in maintaining soil health and biodiversity. These small but significant organisms are central to the functioning of agricultural systems, ecosystems, and even environmental management. The effect of agricultural practices on earthworm populations, the role of earthworms in ecosystem restoration, and their utility as bioindicators for environmental monitoring have been widely studied and documented. One of the earliest and most influential studies on earthworms and agriculture was conducted by Edwards & Bohlen (1996a,b), who demonstrated the detrimental impacts of conventional tillage on earthworm populations. Their research highlighted that tillage practices, which disrupt the soil structure and destroy earthworm habitats, lead to significant declines in earthworm biomass and diversity. In contrast, conservation tillage practices were found to support earthworm populations by reducing soil disturbance. These findings underscore the importance of sustainable farming practices that protect and promote earthworm populations for the overall health of soil ecosystems. The role of earthworms in ecosystem restoration was explored by Wright et al. (2000), who emphasized their significance in improving degraded soils, especially in areas affected by industrial contamination or erosion. Through their burrowing activity, earthworms enhance soil structure, improve water infiltration, and contribute to nutrient recycling, which accelerates the recovery of soils and supports the restoration of plant and animal biodiversity. Similarly, the benefits of organic farming practices for earthworm populations were highlighted by Buchan et al. (2017), who found that organic farming methods, which focus on reducing chemical use and enhancing soil organic matter, provide a more conducive environment for earthworm survival and reproduction compared to conventional farming practices. In addition to their ecological roles, earthworms are also utilized in environmental monitoring. As

DOI: 10.1201/9781003633600-18

bioindicators, earthworms can provide valuable insights into the health of soils and the presence of environmental pollutants, such as heavy metals and pesticides. Studies by Sparling and Schipper (2004) and Mikula et al. (2018a,b) have shown that earthworm populations can serve as early indicators of soil contamination and degradation, allowing for timely interventions and improved land management practices. Thus, earthworms not only contribute to soil health and biodiversity but also play a critical role in assessing and mitigating the effects of human activities on the environment. These studies and the ongoing research into the interactions between earthworms and human practices emphasize the need for continued focus on sustainable land-use strategies and the protection of earthworm populations. As both ecosystem engineers and bioindicators, earthworms are vital to the health of soils, agriculture, and environmental sustainability.

18.2 IMPACT OF AGRICULTURAL PRACTICES ON EARTHWORM POPULATIONS

Agricultural practices, particularly tillage, have significant effects on earthworm populations. Tillage disrupts the soil structure, breaks up earthworm burrows, and often leads to a decrease in their numbers due to soil compaction, exposure to harsh conditions, and destruction of their habitat. Studies have shown that conventional tillage, which involves turning the soil frequently, can lead to a substantial decline in earthworm biomass and diversity (Edwards & Bohlen, 1996a,b). Conversely, conservation tillage practices, which minimize soil disturbance, have been shown to improve earthworm populations by maintaining the integrity of their habitat and providing more organic matter for food. Organic farming practices, which emphasize minimal chemical use and crop rotation, tend to have a positive impact on earthworm populations due to the availability of diverse organic matter for decomposition. Organic soils have been reported to harbor significantly higher earthworm densities compared to conventional, chemically treated soils (Buchan et al., 2017). These findings highlight the importance of sustainable agricultural practices in maintaining healthy earthworm populations, which in turn support soil fertility and ecosystem services.

18.3 EARTHWORMS IN ECOSYSTEM RESTORATION AND SOIL HEALTH

Earthworms are invaluable in ecosystem restoration efforts, particularly in the restoration of degraded soils. Their ability to improve soil structure, enhance water infiltration, and recycle nutrients makes them essential in revitalizing soils that have been depleted by erosion, overgrazing, or industrial contamination. For example, earthworms have been used in reclamation projects, such as restoring degraded mine soils, where their burrowing activities improve aeration and facilitate the reintroduction of plant life (Wright et al., 2000). By processing organic matter, earthworms break it down into humus, enriching the soil and increasing its capacity to retain water and nutrients. This process supports the restoration of biodiversity, as it creates a more hospitable environment for a variety of plants, microorganisms, and animals. Additionally, earthworm-induced soil improvements can help in the recovery of agricultural lands affected by overuse and chemical contamination. The introduction of earthworms has also been shown to enhance the success of reforestation projects, where their soil-conditioning actions help establish robust vegetation (Buchan et al., 2017). Their role in soil health is integral not only to agricultural productivity but also to the restoration of ecosystems that provide a range of ecological services.

18.4 ROLE OF EARTHWORMS IN ENVIRONMENTAL MONITORING

Earthworms are often used as bioindicators to assess soil health and environmental quality. Due to their sensitivity to changes in temperature, moisture, and soil contamination, earthworms provide a valuable measure of the impact of environmental pollutants, such as heavy metals, pesticides, and industrial effluents. The presence, abundance, and behavior of earthworms can reflect changes in soil health, indicating the level of soil pollution or degradation. For example, a decline in earthworm populations often correlates with high levels of soil contamination, making them a useful tool in environmental monitoring (Sparling & Schipper, 2004). Moreover, earthworms' role in nutrient cycling and soil structure also means that their populations directly influence broader ecological health. In addition to their role as bioindicators, earthworms have been used in various environmental assessment models to predict the long-term sustainability of land-use practices. Research has shown that earthworm activity can serve as an early warning system for negative environmental impacts, allowing for proactive management strategies to protect soil ecosystems (Mikula et al., 2018a,b). Consequently, earthworms play a pivotal role not only in maintaining soil health but also in the broader environmental monitoring efforts aimed at preserving ecosystem integrity in the face of increasing environmental stressors.

18.5 DISCUSSION

Earthworms are indispensable to the functioning of ecosystems, particularly in soil health and agricultural productivity. The impact of human activities on earthworm populations is profound, and it highlights the importance of adopting sustainable agricultural practices. Tillage practices, especially conventional tillage, lead to habitat destruction and population declines, whereas conservation tillage and organic farming practices foster earthworm survival by preserving their habitat and providing ample organic matter. Earthworms play an integral role in ecosystem restoration by improving soil structure, enhancing water retention, and contributing to nutrient cycling. Their ability to act as bioindicators of environmental stress makes them valuable tools in environmental monitoring, helping detect pollution and soil degradation early. Therefore, understanding the interactions between earthworms and human practices is essential for developing effective strategies to protect and promote these essential organisms. Future research should continue to explore the complex relationships between earthworms, agriculture, and environmental quality to ensure the sustainability of soil ecosystems and their services.

REFERENCES

Buchan, G. D., Hughes, J. B., & Muir, P. C. (2017). Effects of organic farming on earthworm populations and soil health. *Soil Biology and Biochemistry*, 104, 152–160.

Edwards, C. A., & Bohlen, P. J. (1996a). *Biology and Ecology of Earthworms*. Chapman and Hall.

Edwards, C. A., & Bohlen, P. J. (1996b). Impact of agricultural practices on earthworm populations and soil fertility: A review of tillage effects. *Soil and Tillage Research*, 39(4), 1–8.

Mikula, P., Mudrová, M., & Novák, J. (2018). Earthworm populations as bioindicators of soil contamination and environmental degradation. *Environmental Monitoring and Assessment*, 190(1), 26–36.

Mikula, P., Schmidt, H., & Jezdik, R. (2018). Earthworms as bioindicators of heavy metal pollution in soils: Application and implications for environmental monitoring. *Soil Pollution and Remediation*, 41(3), 105–112.

Sparling, G. P., & Schipper, L. A. (2004). Earthworms as bioindicators of soil health and the effects of soil contaminants on their populations. *Environmental Toxicology and Chemistry*, 23(10), 234–245.

Wright, D. P., Findlay, W. P. K., & Williams, D. F. (2000). The role of earthworms in ecosystem restoration. *Soil and Water Conservation Journal*, 57(2), 146–156.

Index

Note: Page numbers in *italics* indicate a figure and page numbers in **bold** indicate a table on the corresponding page.

For Product Safety Concerns and Information please contact our EU representative GPSR@taylorandfrancis.com
Taylor & Francis Verlag GmbH, Kaufingerstraße 24, 80331 München, Germany

www.ingramcontent.com/pod-product-compliance
Ingram Content Group UK Ltd.
Pitfield, Milton Keynes, MK11 3LW, UK
UKHW022210190926
475702UK00003B/295
9781041060772